THE WESTERN ONTARIO SERIES IN PHILOSOPHY OF SCIENCE

A SERIES OF BOOKS IN PHILOSOPHY OF SCIENCE, METHODOLOGY, EPISTEMOLOGY, LOGIC, HISTORY OF SCIENCE, AND RELATED FIELDS

VOLUME 81

More information about this series at http://www.springer.com/series/6686

Marcus P. Adams • Zvi Biener • Uljana Feest
Jacqueline A. Sullivan
Editors

Eppur si muove: Doing History and Philosophy of Science with Peter Machamer

A Collection of Essays in Honor of Peter Machamer

Springer

Editors
Marcus P. Adams
Department of Philosophy
University at Albany SUNY
Albany, NY, USA

Zvi Biener
Department of Philosophy
University of Cincinnati
Cincinnati, OH, USA

Uljana Feest
Institute of Philosophy
Leibniz University of Hannover
Hannover, Germany

Jacqueline A. Sullivan
Department of Philosophy, Rotman Institute of Philosophy
University of Western Ontario
London, ON, Canada

ISSN 1566-659X ISSN 2215-1974 (electronic)
The Western Ontario Series in Philosophy of Science
ISBN 978-3-319-52766-6 ISBN 978-3-319-52768-0 (eBook)
DOI 10.1007/978-3-319-52768-0

Library of Congress Control Number: 2017933282

Printed on acid-free paper

This Springer imprint is published by Springer Nature
The registered company is Springer International Publishing AG
The registered company address is: Gewerbestrasse 11, 6330 Cham, Switzerland

Contents

Introduction

This volume celebrates Peter Machamer by collecting the work of scholars who, at one time or another, were his students. It is a testament to his intellectual reach, as well as his flexibility and influence as a mentor. All of us—though separated in age by decades and writing across a wide array of subjects and styles—continue to write with his grumbling voice echoing in the back of our minds. We still imagine our latest products returned to us (often within days) with barely legible (and often foul-mouthed) marginalia, pushing us to clarify, elaborate, and (in many cases) simply cut out that [expletive]. For this reason, the volume is subtitled "Doing History and Philosophy of Science with Peter Machamer." Even essays that do not directly engage with Machamer's publications bear the marks of his mentorship, a tough love that was both harsh and eminently supportive.

The volume's wide array of subjects and styles also highlights the flux and diversity of the field of history and philosophy of science and Machamer's role in some of its transformations. For this reason, the volume is titled (tongue-in-cheek) *Eppur si muove*, thereby acknowledging Machamer's work on Galileo, in particular, and alluding to the many ways in which he has moved the field, in general. Machamer has also stressed to us, his students, that *we* were responsible for moving the field. In fact, among his outstanding qualities as a teacher are his acute sensibility for novel and original topics and his willingness to take his students' untutored and sometime heterodox ideas seriously while continuously nudging them to do better and inserting at crucial moments his unique perspective into their projects. It is no coincidence that some of the most influential contributions to recent movements in philosophy of science—concerning mechanistic explanations, values, and inductive risk—were spearheaded by Machamer's students.

In the following, we offer a brief overview of Machamer's academic career and achievements, interspersed with a few observations about how his work has influenced his teaching style. We then offer some remarks about the individual papers collected in this volume.

Peter Machamer earned his earliest renown in the history of philosophy by taking to task Paul Feyerabend's account of Galileo. Feyerabend was then a celebrated, perhaps even revered, scholar, and Machamer a young PhD from the University of

Chicago. With what would become his trademark directness and discernment, Machamer dismantled Feyerabend's view of Galileo as a free-wheeling opportunist and offered instead an interpretation of Galileo as a careful observer and reasoner. Over the next few decades, Machamer continued to elaborate his picture of Galileo: Galileo attempted to explicate the world in terms of a physically intelligible and mathematically tractable mechanism—the balance. In doing so, he exploited elements of the Archimedean-influenced mixed-mathematical tradition, stressing the importance of causal reasoning and mathematical idealization, but returning always to the primary exemplar of the balance. Machamer's findings are now standard in Galileo scholarship. They are represented in his influential 1978 paper "Galileo and the Causes" and, more recently, in "Galileo's Machines, His Mathematics, and His Experiments," which appeared in *The Cambridge Companion to Galileo*, also edited by Machamer.

But Machamer has never been narrowly focused, as his engagement with the variety of trends in early modern science and philosophy clearly demonstrates. Recently, his *Descartes's Changing Mind*, coauthored with J. E. McGuire (Princeton University Press), made a splash among Descartes scholars. In it, Machamer and McGuire argue that Descartes's basic positions shifted dramatically during his life. With verve, they show that the assumption that "Cartesian Philosophy" means *The Meditations* is flawed and that we have much to learn from the evolution of Descartes's views. Importantly, they reveal that the mature Descartes holds an "epistemological stance" on which human knowledge is severely limited and granted, by God, for limited purposes. The impact of this view on the nature of rationalism and speculative philosophy has yet to be fully appreciated, but reviews of the book have been uniformly laudatory. Most repeat a motif often heard in regard to Machamer's work: that it is refreshingly penetrating, that it cuts through complex issues with a deft hand, and that it invigorates positions and debates that others have thought stale.

Machamer's work on Galileo, Descartes, and the mechanical philosophy more broadly is complemented by equally penetrating studies of Hobbes. Machamer has long argued that the consistency and unity of Hobbes's philosophy are underappreciated; indeed, Hobbes may be the most consistent seventeenth-century philosopher, and his system may be the period's most coherent. Machamer's approach to finding unity in Hobbes's philosophy has relied upon examining Hobbes's nominalism and exploiting it to see connections between Hobbes's accounts of language, natural philosophy, politics, and religion. He has recently edited a special issue of *Hobbes Studies* devoted to papers that explore these connections between Hobbes's natural philosophy and other areas.

Machamer's breadth in the history of philosophy is perhaps best exemplified by the wide range of historical dissertations he has supervised. These include works on Roger Bacon, Galileo, Gilbert, Harvey, Descartes, Hobbes, Spinoza, Euler, Auguste Comte, Freud, and Malinowski.

In addition, a word on Machamer's method is appropriate. Machamer is neither a philosopher's philosopher nor a historian's historian. His work does not rest in one methodological gully, blind to happenings in the larger intellectual landscape.

Rather, he has always been sensitive to various historical, philosophical, literary, sociological, and cultural concerns, as well as current developments in the sciences. His work allows conclusions from one area of study to influence others, thus weaving a narrative richer and more penetrating than those of more limited approaches. The ability to draw from unorthodox resources has been at the heart of his pioneering (and often controversial) positions—like those in, say, "Galileo and The Causes" and *Descartes's Changing Mind*. He has urged his students to do the same. For example, he urged students with systematic philosophical interests to consider the history of their subject matter and students with more historical interests to consider the contemporary philosophical implications of their views.

Thus, it is not surprising that Peter Machamer—in addition to supervising work on a truly impressive range of topics—has made important contributions not only to the history of early modern philosophy but also to several debates in contemporary philosophy of science concerning scientific knowledge and the nature of scientific explanation. These two topics are intimately related. A primary aim of science is to advance our understanding of the world by providing explanations. Yet, assessing whether a given scientific explanation truly advances our knowledge requires an understanding of, for example, how that explanation came about, what methods were employed in the process of discovery, and whether and how values and theoretical assumptions may have shaped this process. These questions, once again, highlight how Machamer views science as a complex human endeavor that is situated in multiple historical, social, material, and practical contexts.

Machamer's early work, including such articles as "Observation" (1970), "Recent Work on Perception" (1970), and "Understanding Scientific Change" (1975), exemplifies this point. It was a critical response to the logical positivists' distinction between theoretical and observational languages and as such was part of the general pushback against early twentieth-century philosophy of science in the wake of the works by Thomas Kuhn, Norwood Russell Hanson, and others. Insofar as Machamer brought findings from a variety of different areas of inquiry, including psychology, philosophy, and even art history, to bear on the question of the relationship between perception and knowledge, he contributed to the naturalistic approach in the philosophy of science that was gaining in popularity at the time and that challenged, among other things, the possibility of a theory-neutral observation language. More specifically, Machamer claimed that in order to get clear on the epistemological question of the relationship between perception and knowledge, we must carefully consult all relevant areas of inquiry that have something to say about perception. What these various areas of inquiry reveal, he argued, is that our perceptual systems are structured so as to affect the ways in which information is received from the external world. This highlights the notion—central to Machamer's approach—that knowledge is intimately shaped by a variety of factors, including biological predispositions, expectations, theoretical commitments, and values.

In advocating that philosophers draw from other domains of knowledge in order to address philosophical problems, Machamer's research prompted interest in the philosophy of the special sciences—including anthropology, psychology, and sociology—at a time when physics was still the primary target of analysis and inquiry in the philosophy of science. Machamer instructed us by example that if we are to bring the results of the human sciences to bear on philosophical problems, we need to have a proper philosophy of each of these areas of science. To this end, much of his recent work in contemporary philosophy of science has focused on the nature of theory, method, and explanation in the mind-brain sciences (e.g., *Mindscapes* (1997), *Theory and Method in the Neurosciences* (2001)). Other work has focused on clarifying the role values play in shaping inquiry and knowledge production in science more generally and determining what the implications are (e.g., *Science, Values, and Objectivity* 2004).

Machamer's general focus on issues that arise from scientific practice is also reflected in his important contributions to debates about the nature of scientific explanation. His paper with Lindley Darden and Carl Craver entitled "Thinking about Mechanisms" (2000; hereafter MDC) was inspired by the insight that explanatory practices in the sciences often do not conform to well-entrenched philosophical analyses, such as the covering-law model of explanation. It is the most cited paper ever published in the journal *Philosophy of Science*'s 80-year history (the paper has over 1150 citations). MDC introduced the idea that explanations in the biological and neurosciences could best be understood as describing mechanisms, insofar as such explanations identified the entities and activities productive of phenomena to be explained. To take one of their key examples, neural transmission is a phenomenon that is explained by appeal to the depolarization (activity) of a presynaptic neuron (entity), which results from the opening (activity) of ion channels (entities) and leads to the release (activity) of neurotransmitter (entity) into the synaptic cleft (entity) and binding (activity) of neurotransmitter (entity) to receptors (entity) which become activated (activity) on the postsynaptic neuron (entity). As MDC argue, the accuracy of their account is supported by countless similar examples of explanations in the biological and neurosciences. The paper instigated new interest in the nature of explanation and methodology in the biological and neurosciences. Several papers in this volume reflect this influence.

MDC also harkened back to Machamer's work on the mechanical philosophy, insofar as Boyle, Descartes, Galileo, and Hobbes all sought explanations of phenomena that appealed to the entities and activities that brought them about. This demonstrates that Machamer's work, despite its enormous breadth, is marked by some core themes, and it also shows that Machamer's philosophy of science is profoundly influenced by historical investigations. In this vein, his historical scholarship is typically geared toward addressing, stimulating, or provoking topics and discussions in contemporary philosophy of science.

This volume, while certainly not engaging with the entirety of Machamer's scholarly work in a comprehensive fashion, offers various snapshots of the ways in which his legacy can be seen in the works of his students, both because they continue to work on some of his key ideas and because they echo his method: they demonstrate a sensitivity to both contemporary and historical scientific practices and explore their conceptual, social, scientific, or otherwise systematic implications. We briefly account for these contributions here.

Benjamin Goldberg's essay explores a familiar concept from the philosophy of science, underdetermination, in an unfamiliar context, explanation. Underdetermination is usually deployed in the realism debate or in discussions of theory confirmation. Here, instead, Goldberg is concerned with how underdetermination, interpreted as the necessity of background assumptions, can help us understand a specific historical case involving a dispute about explanatory success. Goldberg examines the work of William Harvey, discoverer of the circulation of the blood, and his rejection of materialist modes of explanation in *De generatione animalium* (1651). He articulates the nature of three background assumptions which affect Harvey's conception of (1) *how* to explain, (2) *what* to explain, and (3) the larger *explanatory stakes* of scientific activity.

Following Goldberg's focus on explanation, Maarten van Dyck reassesses Simon Stevin's mechanics by focusing on how Stevin tried to anchor his mathematical demonstrations in the behavior of material instruments. He shows how his views on the relation between *spiegheling* (speculation) and *daet* (practice) are crucial in correctly understanding his famous proof of the law of the inclined plane and his experimental test of the Aristotelian law of free fall. Van Dyck shows that the distinction between *spiegheling* and *daet* is reproduced in that between instruments at rest and instruments in motion, because of Stevin's claim that impediments to motion are "inseparable accidents" of all moving objects.

David Marshall Miller explicitly engages with Peter Machamer's and Andrea Woody's notion of a *model of intelligibility* (MOI), a concrete phenomenon that guides scientific understanding of problematic cases. His paper extends Machamer and Woody's analysis by elaborating the semantic function of MOIs. He argues that MOIs are physical embodiments of theoretical representations. Therefore, they eliminate the interpretive distance between theory and phenomena, creating classes of concrete referents for theoretical concepts. Meanwhile, MOIs also provide evidence for historical analyses of concepts, like "body" or "motion," that are otherwise thought to be too basic for explicit explication. He illustrates these points with two examples from Galileo. First, he shows how the introduction of the balance as an MOI led Galileo to reject the Aristotelian conception of elemental natures. Second, he shows that Galileo's rejection of medieval MOIs of circular motion constrained the reference of "conserved motion" to curvilinear translations, thereby excluding the rotations that had been previously included in its scope. Both uses of MOIs marked important steps toward modern classical mechanics.

Brian Hepburn expands Miller's treatment of MOIs, focusing on the explication of motion through modeling by simple machines such as the lever and pendulum. He shows that a central way of spelling out the explanatory value of these models is through the concept of equilibrium. Natural motion and simple machines allow the simplification of complex problems in terms of self-evident, intelligible equilibrium conditions. He connects the theme of equilibrium to natural and pendular motion, as well as to mental models in Aristotle, Galileo, Descartes, and Newton. His essay shows that just as equilibrium is a useful model within science, it is also a useful model for doing history and philosophy of science: a normative but objective representation of the important properties of science and its transformation.

Francesca di Poppa turns away from mechanistic explanation to consider a less studied epistemic notion: superstition. She shows that Spinoza connects superstition and piety with the problem of political stability via the notion of obedience. He uses the term "superstitious" to label religious attitudes and practices that promote civil disobedience by establishing demands of allegiance, on the part of the religious authority, that compete with those of the government. Contrary to existing interpretations, di Poppa shows that for Spinoza "superstition" is not characterized by features such as intolerance and anti-intellectualism. In particular, she shows that in the *Theological-Political Treatise*, practices are labeled as pious, rather than superstitious, precisely because they foster obedience and therefore stability, independently of their epistemological valence. She demonstrates how one of Spinoza's goals was to show that what counted as piety among the ancient Jews should be considered superstitious in the more modern, diverse society of the seventeenth-century Dutch Republic and that *libertas philosophandi* was the best way to inoculate a society (if not every single individual) against seditious superstition.

Like other contributors to this volume, Heather Douglas, in a paper entitled "Science, Values, and Citizens," develops a theme she first started working on under the supervision of Peter Machamer and with which she has since opened important new avenues of research: that of the role of values in science. Here, she extends her previous work in considering current debates about public engagement with science, arguing that the central goal of science literacy should not be an understanding of specific scientific facts, but rather of science as an ongoing process of evidence gathering, discovery, contestation, and criticism. Citizens, she argues, should feel that their voice is heard in this process. She distinguishes between various ways in which citizens can engage with the scientific process, specifically honing in on the contestation of research on the basis of values.

In general terms, we can situate the topic of Douglas's article within a broader current in history and philosophy of science, one that looks at the role of science in a liberal democracy and at the role of philosophers as reflecting on science in a political and value-laden context. Warren Schmaus, in his article, "Political Philosophy of Science in Nineteenth-Century France: From Comte's Positivism to Renouvier's Conventionalism," also makes a contribution to this literature, aiming to clarify what the expression "political philosophy of science" might mean. Drawing on a number of historical and contemporary figures (from nineteenth-century French positivism to the more recent works of authors like Kitcher and Longino), he offers a taxonomy of different ways in which philosophy of science can be political. While not endorsing any of these models as the "right" one, he takes his analysis as potentially informing the contemporary debate by offering valuable analytical clarifications.

Schmaus's paper, by virtue of engaging historical material to address a systematic question, exemplifies one of the distinctively Machamerian styles of doing history and philosophy of science identified above. In a similar vein, Uljana Feest's paper, "Physicalism, Introspection, and Psychophysics: The Carnap/Duncker Exchange," takes a debate from the history of early twentieth-century philosophy of science (between the philosopher Rudolf Carnap and the psychologist Karl Duncker) to address the question of the relationship between philosophy and psychology. While this debate has its unique historical context, having to do with the positivist analysis of language as philosophical method, Feest argues that this context may have prevented Rudolf Carnap from seeing that his philosophical tenets were deeply rooted in nineteenth-century psychophysics, a fact that is still largely unknown in contemporary history of analytical philosophy. Feest draws these conclusions in part by looking past Carnap's rhetoric to the kinds of psychological research practices suggested by his argument, as well as to the research practices of then-current psychology.

Jacqueline Sullivan's article, "Long-term potentiation: One Kind or Many?," also makes explicit reference to scientific practice as a driving force of her research, emphasizing that a major impetus she took away from her training with Peter Machamer was his insistence that philosophy of science, if it is to be relevant, should start out not from philosophical fictions about science but from detailed investigations of specific scientific disciplines (in her case, neuroscience) as it is actually practiced. She argues that the experimental practices of neurobiology are not conducive to the discovery of natural kinds as they are sometimes conceived of by philosophers of science. She develops this argument by means of a case study about the history of research on long-term potentiation (LTP).

The issue of practice—this time medical practice—comes up again, albeit in quite a different way, in the article by Thomas Cunningham, titled "Health, Disease, and the Basic Aims of Medicine." In this article, Cunningham also draws importantly on Machamer's work on science and values to argue that the philosophy of medicine can move beyond debates about the true meaning of "disease" and "health" once it is recognized that medicine is value-laden and that patients and clinicians negotiate whether patients' health states are sufficiently dysfunctional to warrant medical intervention.

The last four papers of this volume take as their points of departure the work on mechanistic explanation that was spearheaded by Carl Craver when he was working with Peter Machamer and Lindley Darden. In his article, "Stochastic Supervenience," Craver explores some of the ontic (and, ultimately, metaphysical) commitments underlying the mechanistic position for which he is well known. This position, which found its most prominent exposition in MDC, importantly presupposes an ontic commitment to a deterministic causal structure of the world. Craver raises the question of what the ontic commitments of interlevel mechanistic explanations are, where the explanandum phenomena stand in a supervenience relation to the underlying explanatory mechanisms. Using the possibility of stochastic supervenience as a contrast class and intuition pump, Craver concludes that mechanists are committed to a thesis of strong (i.e., metaphysical, as opposed to merely nomological or physical) supervenience.

Two papers in this volume specifically engage with key aspects of the MDC article that are of particular importance to Peter Machamer, i.e., the concept of activities and the concept of schemata (or schemas). While the characterization of mechanisms as being comprised of entities and activities is well known and cited often, the precise notion of activities remains controversial. In his contribution,

"Activities Are Manifestations of Causal Powers," Gualtiero Piccinini offers a philosophical analysis of this concept that respectfully disagrees with Machamer's own account in "Activities and Causation: The Metaphysics and Epistemology of Mechanisms" (2004). While Machamer proposes an ontology of entities and activities as separate and irreducible to one another, Piccinini argues that activities are properties of entities. More specifically, they are manifestations of causal powers of entities.

Catherine Stinson, in her contribution, entitled "Back to the Cradle: Mechanism Schemata from Piaget to DNA," picks up on the notion of a *mechanism schema*, as it was first introduced in the MDC article, as an abstract representation of a mechanism. In trying to make sense of the notion of a schema, Stinson makes the original move of taking seriously a suggestion made to her (in conversation) by Peter Machamer, namely, that his intended usage of the term "schema" is similar to the way in which the term was used by the developmental psychologist Jean Piaget. Talking about children's cognitive development, Piaget had argued that cognitive schemata, while reflecting previous experience, are tried out, adapted, and modified in response to novel experiential circumstances. This way of understanding the concept, Stinson argues, can serve to elucidate the research process better than other interpretations, which have tended to treat schemas as a mere stage on the way toward more detailed knowledge of a given mechanism. In the second part of her paper, Stinson goes on to provide an account of how abstract schemata can be explanatory.

Like the previous two articles, Tom Seppalainen's paper, "MOCing Framework for Local Reduction," also engages with Peter Machamer's work on mechanisms, focusing on the way in which it can contribute to an analysis of vision research. He critically discusses the concept of a "linking proposition," introduced by the vision scientist Davida Teller (1938–2011). Such propositions express hypotheses about relations of isomorphism between neural and experiential states and are (according to Teller) a crucial part of a reductionistic research strategy within vision science. Seppalainen argues that Teller's model of research is descriptively and normatively inadequate when it comes to research on color vision. Specifically, he argues (a) that the integration of the psychological and the neurological domain occurs through a unifying concept of color processing, called "cancelation," and (b) that Peter Machamer's mechanistic approach, specifically his notion of *activity types*, provides the conceptual resources for a more promising analysis of research in vision science.

Albany, NY, USA — Marcus P. Adams
Cincinnati, OH, USA — Zvi Biener
Hannover, Germany — Uljana Feest
London, ON, Canada — Jacqueline A. Sullivan

Chapter 1
William Harvey's Rejection of Materialism: Underdetermination and Explanation in Historical Context

Benjamin Goldberg

Abstract This essay explores a familiar concept from the philosophy of science—underdetermination—in an unfamiliar context: explanation. Underdetermination is usually deployed in the realism debate, or in discussions of theory confirmation. Here, instead, I am concerned with how underdetermination, interpreted as the necessity of background assumptions, can help us understand a specific historical case involving a dispute about explanatory success. In particular, I look at the work of William Harvey, discoverer of the circulation of the blood, and his rejection of materialist modes of explanation in the course of his *De generatione animalium* (1651). I articulate the nature of three background assumptions at work here, which affect Harvey's conception of: (1) *how* to explain; (2) *what* to explain; and (3) the larger *explanatory stakes*.

Underdetermination is a contested idea in philosophy of science, one with a large (and confusing) body of literature.[1] However, my dance here is different from the usual philosophy of science two-step: I won't argue for underdetermination's

[1] The following articles are a small selection of those that have been most important in my own understanding of the debate: Bogen, James and James Woodward (1988), "Saving the Phenomena," *The Philosophical Review*, 97(3): 303–351; Doppelt, Gerald (2007), "The Value Ladeness of Scientific Knowledge," In: *Value-Free Science? Ideals and Illusions*. Eds. Harold Kincaid, John Dupré, and Alison Wylie. Oxford: Oxford University Press; Kincaid, Harold (2007), "Contextualist Morals and Science," In: *Value-Free Science? Ideals and Illusions*; Magnus, P.D. (2005a), "Hormone Research as an exemplar of underdetermination," *Studies in History and Philosophy of Biology and Biomedical Sciences*, 36: 559–567; Magnus, P.D. (2005b). "Reckoning the shape of everything: Underdetermination and cosmotopology," *British Journal for the Philosophy of Science*, 56(3): 541–557; Okasha, Samir (2000), "Underdetermination and the Strong Programme." *International Studies in the Philosophy of Science*, 14(3): 283–297); Okasha, Samir (1997), "Laudan and Leplin on Empirical Equivalence," *British Journal for the Philosophy of Science*, 48: 251–256; Roush, Sherrilyn (2007), "Constructive Empiricism and the Role of Social Values in Science," In: *Value-Free Science? Ideals and Illusions*; and Ruphy, Stéphanie (2006), "Empiricism

B. Goldberg (✉)
Department of Humanities and Cultural Studies, University of South Florida, Tampa, FL, USA
e-mail: big@usf.edu

M.P. Adams et al. (eds.), *Eppur si muove: Doing History and Philosophy of Science with Peter Machamer*, The Western Ontario Series in Philosophy of Science 81, DOI 10.1007/978-3-319-52768-0_1

existence, nor for any far reaching anti-realist consequences. Instead, in honor of my dear teacher and friend Peter Machamer, who always encouraged me to try the more complicated routines of a properly integrated history and philosophy of science, and to move beyond lazy antiquarianism to something, as he would put it, "actually interesting,"[2] One of the lessons that I learned from Peter was to distrust philosopher's accounts of science: he argued that actual science was almost always more complicated than our reconstructions, and he would often excoriate, in blistering language, problems like scientific realism as pseudo-problems, based upon overly simplistic ideas of scientific practice and theory. But he was quick to point out that, as Lakatos had argued, history of science without philosophy of science was blind (Lakatos 1970).[3] Here, then, I hope to combine history with philosophy, and explore how a standard and oft-gnawed chestnut of philosophical dispute, *underdetermination*, is useful for understanding another such debated idea, *explanatory success.* I apply this theory to a particular case from the history of science in order to appreciate the true nature and complexity of explanatory success. In particular, I focus on the early modern context of the great physician and philosopher William Harvey's repudiation of materialist modes of explanation in his final published work, the *Exercitationes de generatione animalium* (1651).

I begin by briefly outlining underdetermination, before describing Harvey's discussion of explanation. I then articulate the nature of three background assumptions at work here, assumptions that affect: (1) *how* to explain; (2) *what* to explain; and (3) the *explanatory stakes*. In the fourth section, I conclude with some thoughts about some large-scale changes in natural philosophy witnessed over the course of the seventeenth century.

1.1 Underdetermination and Background Assumptions

Underdetermination is often interpreted to be a claim about the existence of empirically equivalent rivals. Helen Longino, however, has offered a different account, one that has not always explicitly recognized in the current literature.[4] Indeed, Longino has been consistently misread on this issue, and has been interpreted as offering a relatively standard thesis about empirically equivalent rivals; Kinkaid, for example,

---

all the way down": a defense of the value-neutrality of science in response to Helen Longino's contextual empiricism," *Perspectives on Science*, 14 (2): 189–214.

[2] In fact, Peter would (and did) use a lot more expletives when attacking my antiquarianism; he was never one to bandy about with words, and I have always deeply appreciated (and benefited from) his plainspoken honesty.

[3] This is not to say that Peter or I are Lakatosians. Peter's allegiance is, in fact, hard to pin down, if he even has one, if only because he has read—and *learned* from—just about everyone writing on whatever issue is under discussion, from wine to Wittgenstein. I recall the intimidatingly high stacks of *complete* books Peter had stacked up around his house with both fondness and envy at his ability to absorb books as if through osmosis.

[4] One exception is Ruphy 2006.

writes of Longino's underdetermination thesis that, "Comparing 'all the data' with 'all the possible hypotheses' is an incoherent situation we are never in."[5] This is not an accurate interpretation: Longino understands underdetermination as the thesis that data cannot be used as evidence for a theory without additional assumptions. While this sounds similar to the standard account of Duhem-Quine holism and underdetermination, it is importantly different in that Longino's account focuses not upon how data is confirmatory, but something logically prior: the *interpretation* of data *as* evidence.

Longino's most basic description of underdetermination is in terms of a "semantic gap between statements describing data and hypotheses[.]"[6] A 'semantic gap' occurs when there are theoretical terms used in our theories (e.g., electrons) that are not contained in our descriptions of data (e.g., tracks in a cloud chamber). Longino links this semantic gap to the idea that, "Data—even as represented in descriptions of observations and experimental results—do not on their own… indicate that for which (hypotheses) they can serve as evidence."[7] Background assumptions are what might also be called 'auxiliary hypotheses'[8] or even 'enthymematic premises'.[9] Indeed, similar ideas (though, as I'll remark in the conclusion, different in important ways) are often used to help understand the notion that certain complexes of ideas are 'in the air' during certain historical periods, such as Whitehead's 'climates of opinion,' Kuhn's 'paradigms,' or Dear's notion of 'intelligibility'.[10] Whatever they're called, what I term background assumptions consist of a variety of different sorts of things: accepted theories, empirical observations, statistical and mathematical axioms, methodological imperatives, tacit knowledge, physical laws, broad metaphysical theses, and so on.

Longino's claim is that data or observations are unable to indicate what they are evidence for *by themselves*: they require an interpretation in order to *become* evidence. There must be a connection between however the data are described—as a set of observation sentences, as the output of some experiment that results in a set of numerical values for some set of variables, etc.—to an aspect of the theory or

[5] Kincaid 2007, 222.

[6] Longino, Helen (2002), *The Fate of Knowledge*. Princeton: Princeton University Press, 50; see also, Longino, Helen (1990), *Science as Social Knowledge*. Princeton: Princeton University Press, 52

[7] Longino 1990, 58.

[8] Feyerabend, P.K. (1965), "On the 'Meaning' of Scientific Terms," *The Journal of Philosophy*, 62.10: 266–274.

[9] This goes back to Aristotle's idea of an incomplete syllogism, but more recently Alan Gross has used this notion of enthymematic reasoning in his analyses of public science and science policy; see: Gross, A. G. (1994a), "Is a rhetoric of science policy possible?" *Social Epistemology* 8: 273–80, and Gross, A. G. (1994b), "The roles of rhetoric in the public understanding of science," *Public Understanding of Science* 3: 3–23.

[10] Whitehead, A.N. (1953), *Science and the Modern World*, New York: The Free Press, 17. I discuss Kuhn in the conclusion. Peter Dear's notion of intelligibility is very close to the picture I lay out here; see: Dear, Peter (2008), *The intelligibility of nature: How science makes sense of the world*, Chicago: University of Chicago Press.

hypothesis or hypotheses for which it they are supposed to stand as evidence. Let me illustrate this by way of a toy example.[11] Suppose I am driving down a snowy Pittsburgh street on a winter evening, and suddenly a naked person runs in front of my car.[12] For what hypothesis does this event stand as evidence? The answer depends upon what sorts of assumptions I have that are relevant to the situation. If I have no background assumptions that allow me to connect this event to any claim other than 'some naked people just ran by,' then this event doesn't serve as evidence for any other claim at the time, though, of course, it may serve as evidence at some point in the future. If, on the other hand, we suppose that I have the assumption that there is a fraternity in the neighborhood I am driving through, then the naked runner serves as evidence for the claim that, 'some fraternity members may have over indulged,' provided I have other relevant assumptions, such as 'only drunk fraternity members run naked in winter,' and so on. On the other hand, suppose I have the assumptions that the Steelers were competing that night, and that a victory would cause fans to celebrate in strange ways; if these are my assumptions, then the naked runner serves as evidence for the claim that, 'the Steelers have won the game.' Thus, in cases where I hold different background assumptions, the same state of affairs (the naked runner) serves as evidence for different claims (drunken college students vs. a Steeler's victory). Interpretation depends upon background assumptions:

> …how one determines evidential relevance, why one takes some state of affairs as evidence for one hypothesis rather than for another, depends on one's other beliefs, which we can call background beliefs or assumptions. Thus, a given state of affairs can be taken as evidence for the same hypothesis in light of differing background beliefs, and it can be taken as evidence for quite different and even conflicting hypothesis given appropriately conflicting background beliefs.[13]

Background assumptions form preconditions that are needed for evidential reasoning.[14]

So far, so good. What I want to suggest now is that something broadly similar is at work when it comes to explanation. Background assumptions about, for instance, the nature of living beings, about scientific methodology, and about explanation itself, are needed in order to connect explanatory hypotheses with whatever set of phenomena or observations they are meant to explain. Indeed, this might be seen to follow trivially insofar as theories are explanations of the interpreted data that serve as their evidence: the semantic gap happens both ways.[15] Longino's semantic gap

---

[11] The following example is inspired by one in Longino 1990.

[12] Don't worry, I missed them; I'm an excellent driver.

[13] Longino 1990, 43.

[14] This is not to claim, however, that all background assumptions are equal or are needed to the same extent; nor that assumptions are themselves immune to confirmation or are always non-empirical; nor is it to claim that background assumptions do not include data themselves, or rather, interpreted data.

[15] Of course, theories are meant to explain all possible relevant data and observations, but this of course includes whatever data or observations were used in the construction or confirmation of that theory.

structures not just how we interpret data as evidence for theories, but also how we interpret theories as explanatory. Since background assumptions affect our descriptions of data and observations such that they can be connected with particular theories, different ways of describing and interpreting data necessarily affect what features stand in need of explanation, and how those explanations are supposed to function.

My point here is simple and (hopefully) unobjectionable. I do not mean to establish any broadly (or even narrowly) antirealist theses about the nature of science, or to establish any particular account of scientific explanation.[16] Instead, given underdetermination, disagreements about scientific explanations don't necessarily indicate disagreements about evidence. Instead, they might indicate much more fundamental disagreements about various, recondite background assumptions, disagreements about what features of the world need to be explained, and how that explaining needs to be accomplished: assumptions structure how scientists think about explanatory success. Indeed, from this perspective, we can see that underdetermination demonstrates the profound relevance (and location) of philosophical disagreement in science. So, to return to the above example, if a friend disagreed about, for instance, the assumption that only drunken fraternity members would run naked through the snow, she might also disagree with my explanation: she might suggest, for instance, that it could have been a bet gone wrong, or simply an instance of a profoundly stoic person training themselves under adverse conditions. Indeed, depending upon her assumptions, she might even reject the need to explain naked running entirely.[17]

To learn about scientific disagreement, then, we can begin by articulating the background assumptions at issue in these instances of explanatory incongruence. In the remainder of this paper, I attempt to flesh out this idea in the context of early modern philosophy and medicine.

1.2 William Harvey on Materialist Explanations

William Harvey is best known for his *De motu cordis*,[18] an argumentative tour-de-force that established that the blood moves around the body in a circle, forced along by the forceful systole that ejects blood from the heart. Harvey's last published work, the *Exercitationes de generatione animalium*,[19] is a very different work, and

[16] For some rumination on the latter, however, see: McIntyre, L. C. (2003). "Taking underdetermination seriously." *SATS: Nordic Journal of Philosophy, 4*(1), 59–72.

[17] This may not sound believable, but as an undergraduate at Carleton College in Northfield, MN, I learned that streaking could be a favored hobby for many, and in every season. I eventually stopped asking why some naked individual was jogging (occasionally even walking) across campus.

[18] Harvey, William 1628, *Exercitatio anatomica de motu cordis et sanguinis in animalibus*, Frankfurt.

[19] Harvey, William 1651, *Exercitationes de generatione animalium*, London.

while it was widely read and appreciated in its time, it has not fared nearly so well among modern historians as *De motu,* at least until recently.[20] While early historians of science used Harvey's *De motu* to place him in the pantheon of scientific revolutionaries and new mechanical philosophers, more recent scholarship has pointed out the flaws in this assessment, flaws which are especially apparent when we turn our gaze to the *De generatione.*[21] This work demonstrates Harvey's commitment to Aristotle's philosophy, and his rejection of new, materialist modes of explanation. Harvey's work was certainly championed by the mechanical philosophers, but he, himself, did not belong to that club.[22]

Harvey straightforwardly rejects materialist modes of explanation. His repudiation is found in the eleventh exercise of his work on generation, where he outlines his disagreement with the ideas of his teacher in Padua, Fabricius ab Aquapendente (1537–1619). Having established in earlier chapters how an egg is generated in a chicken, Harvey discusses the parts of the egg, starting with the shell. It is in this this context that Harvey comments on materialist explanations about the phenomena of generation, in particular epigenesis, the construction of the embryo, part by part, over time.

> The usual error of those who philosophize these days is to seek the causes of the diversity of the parts from the diverse matter out of which they arise. So the physicians affirm that the different parts of the body are fashioned and nourished from the diverse materials of blood, or of sperm: as everyone knows, from the thinner matter, the soft parts, like the flesh; and from harder and thicker matter, the earthy parts, like bones etc. But I have refuted this exceedingly widespread error elsewhere.[23] Equally deceived are those who make all things from atoms, like Democritus, or from elements, like Empedocles. As if generation were nothing other than the separation, or the collection, or the arrangement of things. I do not indeed deny that in order for one thing to be produced from another, all these aforementioned things are necessarily required; but generation is itself distinct from all of them. I find Aristotle to be of this opinion, and I will myself hereafter show that from the same white of the egg (which everyone admits to be a similar body and not composed of diverse parts) every part of a chicken (bones, claws, feathers, flesh and all the rest) is produced and nourished. Moreover, those who philosophize in this manner assign only a material cause, and deduce the causes of natural things either from a concurrence of the elements happening by design or by chance, or from diverse arrangements of atoms. They do not touch on

[20] See especially: Lennox, James 2006, "The Comparative Study of Animal Development: William Harvey's Aristotelianism," In: *The Problem of Animal Generation in Early Modern Philosophy*, Ed. J.E.H. Smith, Cambridge: Cambridge University Press. Although Roger French discusses it in his work on Harvey, he has no chapter on it, unlike *De motu cordis* and even one on Harvey's lecture notes, the *Prelectiones anatomie universalis*. Neither the work on generation itself nor the topic of 'generation' appears in the index; see French's 1994 *William Harvey's Natural Philosophy*, Cambridge: Cambridge University Press; and Harvey, William 1616 [1964], *Prelectiones anatomie universalis*, Ed. Gweneth Whitteridge, London: Royal College of Physicians.

[21] Lennox 2006.

[22] At least in the traditional sense of 'mechanical philosopher'. For a much better way of thinking through issues of mechanism in the context of early modern medicine in general and Harvey in particular, see: Distelzweig, Peter (2016), "'Mechanics' and Mechanism in William Harvey's Anatomy: Varieties and Limits," In: *Early Modern Medicine and Natural Philosophy,* Eds. Peter Distelzweig, Benjamin Goldberg, and Evan Ragland, Dordrecht: Springer.

[23] C.f. Harvey 1651, Ex.45 and 72.

> that which is special in the operations of Nature, and in the generation and nutrition of animals: for they do not recognize the existence of the divine Agent and the deity of Nature (who works with the highest skill, foresight and wisdom, and who produces all things to some certain end or for the sake of some certain good). They derogate from the honor of the divine Architect, who made the shell to be the guardian of the egg with no less skill and foresight than he composed all the rest of the parts of the egg out of the same material and through the same formative power.[24]

In the following section, I analyze this argument in terms of its background assumptions. We shall see that Harvey's rejection of explaining epigenesis in terms of material causes depends upon a host of substantive, often metaphysical, background assumptions about both the nature of living beings and the proper modes of scientific investigation and explanation proper to them.

Before moving on, one last piece of stage setting: answering the question of why it's important to understand Harvey in terms of underdetermination.[25] One important answer, though by no means the only one, is that understanding Harvey's argument about explanatory strategies concerning generation (and, more generally, living things) serves to emphasize that these questions were *open* and *under debate*. Some historiography of philosophy, especially that concerned with canonical modern philosophers like Descartes, tends to treat debates over method with Aristotelian and other classical philosophies as having forgone conclusions, a sort of mild anachronism that discounts the arguments of Scholastics as reactionary, as being on the wrong side of history.[26] Indeed, for the most part, the study of early modern Aristotelianism has been on the fringes of history of philosophy scholarship. Part of what makes Harvey so interesting is that his discoveries—trumpeted by Descartes, Hobbes, and others—were, from his own point of view, intimately related to and based upon an Aristotelian understanding of the world.[27]

1.3 Assumptions About the Nature of Living Things

Before moving on to analyzing what background assumptions are operative, it will be helpful to discuss the structure of Harvey's argument in some detail. First, note that there are two distinct theses that Harvey rejects: the first is the idea that a diversity of parts must be caused by a diversity of matter. The second is what we might

[24] Harvey 1651, Ex. 11, 28–29.

[25] Also note that I do not explicitly argue for underdetermination in this case, instead I assume that underdetermination is a fact about scientific reasoning.

[26] Indeed, recent historiography has demonstrated the extent to which the debate about living things and their proper mode of study and explanation reveals fault lines even within the camp of materialist, mechanical philosophers. For instance, see: Easton, Patricia and Melissa Gholamnejad (2016), "Louis de la Forge and the Development of Cartesian Medical Philosophy," In: *Early Modern Medicine and Natural Philosophy*, Eds. Peter Distelzweig, Benjamin Goldberg, and Evan Ragland, Dordrecht: Springer.

[27] Though, it must be said, a decidedly Renaissance and eclectic version of Aristotelianism.

call the atomic or corpuscular hypothesis, the idea that all things are composed of (and *only of*) atoms or elements. The 'only' part is quite important because, as is obvious from any examination of his work, Harvey clearly believes that elements and matter are important for understanding living bodies.

Second, in between these two theses, Harvey interposes an example of the first thesis he is rejecting ("...the physicians affirm that the different parts of the body are fashioned and nourished from the diverse materials of blood, or of sperm..."). What is important here is that Harvey believes he has refuted this idea *on empirical grounds,* which he relates in Exercises 45 and 72. So an argument for the first thesis, that a diversity of parts requires a diversity of matter, is straightaway rejected on the basis of his observation that it is from an entirely *homogeneous* mass that the various, heterogeneous parts of the developing chick are made.

Third, the rest of Harvey's argument has a three-part structure. First, he restates the thesis he is rejecting: the idea that generation is accomplished through the addition, subtraction, or rearrangement of matter. The first part of Harvey's argument states that, while these processes are necessary and important, generation is not *reducible* to these processes alone. This is quite important, because it helps us understand Harvey's relationship to materialist ideas. Importantly, he doesn't reject materialist explanations *tout court*, but only explanations appealing to a reduction to matter alone.[28] Indeed, such a finding should perhaps temper our enthusiasm for oppositions between large-scale historical categories, like materialism vs. animism, or mechanism vs. vitalism. So, for instance, as Charles Wolfe cogently argues, there are many distinct kinds of materialism, including a kind of vital materialism that attempts to make coherent the desire to understand phenomena in terms of their material components, while, at the same time, respecting the self-organizing power and functionality of living things.[29] Indeed, though his terms and conclusions are far removed from eighteenth century materialists, Harvey, too, is attempting to resolve tensions arising from the desire to respect both these aspects of nature.

The second part of Harvey's argument connects his position with that of his 'general,' Aristotle, a fact that will become relevant below when I turn to examine Harvey's background assumptions.[30] The third and final part gets to the core of his objection to materialist reduction, namely, that such reductions entirely ignore what Harvey takes to be the defining feature of living things: their teleological nature. Harvey's conception of anatomy and medicine is deeply teleological, in terms of both its methods and its metaphysics, and, as I've argued elsewhere, this conception

---

[28] While 'reduction' is not an actor's category, I employ it because it connects with some of the literature on Aristotle's ideas about explanation, noted below. I use it to mean something like 'explanations appealing to matter/elements alone.'

[29] Wolfe, Charles T. (2014), "Materialism," In: *The Routledge Companion to 18th Century Philosophy*, Ed. Aaron Garrett, Routledge: New York.

[30] Harvey, in the Preface to the *De generatione,* calls Aristotle his 'General' and Fabricius his 'guide.'

of science bleeds into natural theology, and so this last part of the argument depends upon the Divine.[31]

With this understanding of Harvey's argument in hand, I will now discuss three deeply interrelated background assumptions: anti-reduction, teleology, and natural theology.[32] I note further, that we can see here three distinct ways by which a background assumption affects explanation, to wit: by affecting *how things are to be explained,* by affecting *what things need to be explained,* and by affecting *the larger philosophical or theological import of the explanatory situation*, the *explanatory stakes*.

I start with Harvey's anti-reductive argument. He writes that a whole host of material processes are needed in order for generation to occur, but that generation is "...itself distinct from all of them." The background assumptions here concern basic strictures of Aristotelian science and modes of explanation concerning natural things that Harvey follows attentively.[33] Aristotle's view of generative processes was based on his metaphysics of actuality and potentiality.[34] In particular, generation is a process wherein a single potential for an organism, the form of the complete living being, is actualized. In this process, element potentials, that is, material causes, are important, but the process of generation cannot be understood or reduced to them. As Allan Gotthelf has argued, "...the irreducibility to element-potentials for organic development is the core of the meaning of the assertion that the development is for the sake of the mature organism, and thus the core of Aristotle's conception of final causality."[35] Importantly, this assumption of teleology informs Harvey's understanding of what it would take to explain the phenomena of generation. Reductive explanations are non-starters since the phenomena of generation are, by definition, not reducible to material-based explanations alone.

Definitions play an important role as background assumptions, but it's essential to realize that for Harvey, as for Aristotle, definitions are often *empirical.* And so it is in this case where Harvey's assumption about the impossibility of a reductive explanation of generation depends upon defining generation (in perfect, blooded animals, at least) as *epigenesis,* that is, as the creation of the offspring part by part, over time. Harvey believes he has experientially verified by performing a series of

[31] Goldberg, Benjamin (2013), "A Dark Business, full of shadows: analogy and theology in William Harvey," *Studies in History and Philosophy of Biological and Biomedical Sciences* 44: 419–432.

[32] Individuation is, as ever, a tough problem. I would be equally happy to say this is one complex background assumption, one that is based in ideas about teleology, but it is easier to tease out important philosophical threads if we separate it into these three distinct parts.

[33] For some details on how just how close Harvey's ideas of generation are to Aristotle's (and they are close indeed), see Lennox 2006; see also Goldberg 2013.

[34] My account here follows Allan Gotthelf's classic paper (1987), "Aristotle's Conception of Final Causality," In: *Philosophical Issues in Aristotle's Biology*, Eds. Allan Gotthelf and James Lennox, Cambridge: Cambridge University Press. See especially: 212–214.

[35] Gotthelf 1987, 213.

observations of chick embryos in eggs over their period of development.[36] This is the sticking point that causes Harvey to reject Fabricius' explanation *ex materia*:

> For this reason Fabricius wrongly sought the material of the chick as some distinct part of the egg out of which it should be embodied, as if the generation of the chick were performed by metamorphosis or by the transfiguration of a mass of material, and all the parts of the body, or at least the principal parts, arose all at the same time and (as he says) were embodied out of the same material, and not by epigenesis in which an order is observed according to the dignity and use of the parts; where first a tiny foundation, as it were, is laid which at the same time as is grows, is divided and shaped and then acquires its parts, which arc engendered from it and born with it, each in its own order.[37]

So given that, empirically, Harvey has defined generation as epigenesis, and given the background assumption about non-reductive explanation, Harvey could not *but* reject purely materialist explanations of generation, since such a rejection was the only option that wouldn't violate either his observations or his theoretical and metaphysical commitments. We see, then, a primary way by which background assumptions affect explanation: they guide the ways in which certain sorts of phenomena are to be characterized and accounted for, and thus Harvey rejects Fabricius' (and other physicians) reductive explanations of generation as simply invoking the wrong sort of *explanantia*, referencing matter and motion where (as we shall see in more detail below) ends and teleology are needed.

Importantly, epigenesis was a sticking point for just about every later materialist philosopher: if matter was passive and inert, devoid of any powers or potentialities for complex, regular change, how could one possibly hope to explain the byzantine, yet orderly, process by which an embryo develops? Reductive explanations seem not to allow for this process to be explicable. As Joseph Glanville opined nearly 20 years after Harvey's death,

> Blind Matter may produce an elegant effect for once, by a great Chance; as the Painter accidently gave the Grace to his Picture, by throwing his Pencil in rage, and disorder upon it; But then constant Uniformities, and Determinations to a kind, can be no Results of unguided Motions.[38]

Matter alone could not account for the dignified, orderly process of epigenesis, and so it was often rejected in favor of preformationist doctrines. Part of the reason why epigenesis proved so problematic for materialist philosophers was that, in addition to its complexity and seemingly self-organizing manner, it seemed to undercut a

[36] For which see Harvey (1651), Exs. 42–45, 113–125). I discuss some of Harvey's ideas on generation, including his attempt at creating an analogy about conception in *brain* and *womb* in Goldberg 2013.

[37] Harvey 1651, Ex.45, 123.

[38] Glanvill, Joseph 1676, *Essays on Several Important Subjects in Philosophy and Religion*, London. In his 1665, *Scepsis Scientifica; or, Confest Ignorance, the Way to Science*: *in an Essay on the Vanity of Dogmatizing and Confident Opinion*, London (of which the *Essays* are an abbreviated form), Glanvill makes a similar comment, and writes that such processes as those which show uniformity and conformity must be, "... regulated by the immediate efficiency of some knowing agent: which whether it be seminal Forms, according to the Platonic Principles, or whatever else we please to suppose; the manner of its working is to us unknown" (39–40).

central tenet of much materialist philosophy, namely their anti-teleological conception of nature. Descartes' system in particular foundered in these waters. It seemed to many early modern thinkers that such explanations would depend upon mere chance or fortune in order to explain such a process, and thus reductive materialist explanations could not work—without, at least, recourse to God's will.[39] So, important for our consideration of Harvey's rejection of materialism, we note that he characterized generation in terms of final causes.

Harvey's assumptions about reduction and epigenesis thus bleed into the next assumption, indeed, can be seen to follow from it: namely, the fundamentally teleological conception and end-oriented viewpoint of Aristotelian science. As is well known, a core debate in natural philosophy across the seventeenth century concerns the basic philosophical account of scientific method and modes of explanation. The traditional account would argue that, starting with Descartes, we see the gradual elimination of teleology from natural philosophy. And while there is some truth to this in accounts of the physical sciences,[40] the story in medicine and the burgeoning life sciences is much more complex, for here teleology seemed a fundamental organizing feature of the world—though there is a shift from *natural* Aristotelian immanent teleology towards an *unnatural* (Christianized) Platonic conception, wherein ends stem from God's Will and Design (about which more below).[41] In any event, final causality is certainly central for Harvey, as teleology is core to his entire conception of anatomical science and method.[42]

This is clearly seen in the quote above, where Harvey argues that 'what is special in the operations of Nature' just is the fact that all natural things operate with respect to some end or for some good. In the special case of animal bodies, Harvey articulates the Aristotelian conception of the body as the instrument of the soul, wherein each part is *for the sake of* some particular biological function or purpose (Harvey often invokes Aristotle's *hou heneka* with the more literal translation of *alicius gratia* instead of the Scholastic *causa finalis*).[43] Indeed, as I've argued elsewhere, teleology is central to Harvey's entire project, and he defines anatomy *just as* the search for certain kinds of teleological causes:

[39] Des Chene, Dennis 2003, "Life after Descartes: Régis on Generation," *Perspectives on Science* 11(4): 410–420.

[40] Though see Jeffery K. McDonough on least action principles and monadic teleology, especially his (2016) "Leibniz on Monadic Teleology and Optimal Form," *Studia Leibnitiana Sonderhaft*, and his 2011 "The heyday of teleology and early modern philosophy," *Midwest Studies in Philosophy* 35.1: 179–204.

[41] The importance of teleology is apparent from Jacques Roger (1963), *Les Sciences de la vie dans le pensée françaese du XVIIIe siècle*, Paris: Armand Colin. For discussion of these two modes of teleology, see: Lennox, James (1985), "Plato's unnatural teleology," *Platonic investigations,* Ed. Dominic O'Meara, Washington: Catholic University of America: 195–218.

[42] Harvey's conception of anatomical science and the importance of teleology in much more detail in Goldberg (Forthcoming), "Anatomy as a Science of Teleology: The Case of William Harvey," In: *Interpretations of Life in Heaven and Earth*, Ed. Hiro Hirai, Dordrecht: Springer. See also French 1994.

[43] E.g., Harvey 1616 [1964]), 22.

> …the end of Anatomy is to know or be acquainted with the parts and to know them through their causes and these [i.e., the causes,] 'that for the sake of which' and 'that on account of which,' [should be known] in every animal…[44]

Indeed, we can directly connect Harvey's conception of the end of anatomy with underdetermination: without this rather substantive assumption about the proper goal of anatomical science, what needs to be explained is not specified. And so we see another important way by which background assumptions affect explanation, namely, by stipulating those things that stand in need of explanation in the first place. Indeed, later generations of anatomists shift away from understanding their discipline as primarily concerned with final causes, and thus a major shift in the life sciences consists not just in the advancement of empirical observation (though that is of course important), but also in the rejection of certain explanatory demands as important or even relevant. So, roughly a century later, we see a new characterization of anatomical science in the work of Albrecht von Haller, who argues that he had to dissect and vivisect in order to investigate the causes of animal motions, and that "All of physiology lies in the internal and external motions of the living body."[45] That is, though understanding function remains a central task of his science, the way in which this concept is articulated takes place in terms of a set of materialist or mechanist background assumptions, where the emphasis is less upon functions *as organizing ends*, and instead upon functions *as motions* and *mechanical causes*. Background assumptions about the nature of the world and living things thus play a substantial role in determining what phenomena are in need of explanation.[46]

It is on exactly this subject of what needs to be explained that the third background assumption becomes important: this assumption helps interprets natural things as not just having specific ends, but also as being *well designed*. Thus part of what's wrong with materialist modes of explanation by Harvey's lights is that, not only do they mistake material causes as sufficient (the first assumption), nor do they just ignore the teleological status of living things and their parts (the second assumption), but that, in so doing, they 'derogate from the Divine Architect.' Harvey has inherited the Renaissance's complex, eclectic philosophical doctrines, doctrines

[44] Harvey 1616, 22. See also: Goldberg (forthcoming), "William Harvey on Anatomy and Experience," *Perspectives on Science* 24:3.

[45] Haller, A von, 1757–1766. *Elementa physiologiae corporis humani*, Vol.1: *Fibra, Vasa, Circuitus sanguinis, Cor*. Lausane, iii. See also Andrew Cunningham (2002), "The pen and the sword: recovering the disciplinary identity of physiology and anatomy before 1800 I: Old physiology—the pen," *Studies in History and Philosophy of Science Part C: Studies in the History and Philosophy of Biology and the Biomedical Sciences*, 33.4: 631–665. Note Harvey uses 'anatomy' and Haller 'physiology'—this is an important difference, but here is not the place to discuss it. See also: Cunningham's (2003), "The pen and the sword: recovering the disciplinary identity of physiology and anatomy before 1800: II: Old anatomy—the sword," *Studies in History and Philosophy of Science Part C: Studies in History and Philosophy of Biological and Biomedical Sciences*, 34(1), 51–76.

[46] Another, perhaps more interesting way of putting this would be to understand background assumptions as not just picking out what needs to be explained, but as actually constructing the image we have of the phenomena *tout court*, following along the lines of Bogen and Woodward 1998.

that amalgamate Plato with Aristotle, and both with (in Harvey's case) a notably ecumenical Christianity.[47] In other words, Harvey's teleology is deeply linked with his natural theology.[48] And so we see that this background assumption affects not just the explanatory situation, but what we might call the *explanatory stakes:* that is, the larger philosophical or theological import of explaining generation. It is obvious that scientific explanation is deeply political and social, that is, it affects our image the world and ourselves in multifaceted ways. The stakes in the life sciences should be obvious to anyone acquainted with debates over creationism and evolution in the United States. The stakes in Harvey's time were no less severe, and it is well known that materialism had long been associated with the specter of atheism; indeed, we can understand Pierre Gassendi's project in part as an attempt to create a theologically responsible conception of corpuscularianism, something taken up wholeheartedly in England by Robert Boyle and others.[49]

Unlike Descartes, who rejected teleology and natural theology, and, indeed, argued in (for example) the Fourth Meditation that one could not have knowledge of intentions of Divine Actions,[50] Harvey's background assumptions about the Divine Design of the world are key to understanding his rejection of materialist modes of explanation: any mode of explanation which reduces a phenomena to chance and concurrence, that eliminates the foresight, wisdom, and skill found in the construction of living beings and the manner of their generation, is not simply shirking its epistemological responsibilities, but is, in fact, doing something theologically suspect. Harvey's background assumption causes him to interpret animal bodies and their parts as *wisely designed*, something he comments on throughout *De generatione.* In fact, this background assumption means that he understands the whole process of generation in just these sorts of terms, for instance, describing the process and substances involved in fertilization as somehow divine: "This is easily seen from its wondrous operations, its contrivance and wisdom in which there is nothing done to no purpose or rashly or by chance, but all things are established for the sake of some good and to some end."[51] Harvey quite literally describes his subject matter not just as being judicious and well designed upon reflection, but as

---

[47] In the quoted passage, Harvey refers variously to the 'divine Agent,' the 'deity of Nature,' and the 'divine Architect.' Throughout the *De generatione,* he refers to the Creator using a variety of (presumably) interchangeable terms.

[48] See the last section of Goldberg (2013), where I connect his use of analogy to his theological views.

[49] Lennox, James (1983), "Robert Boyle's defense of teleological inference in experimental science," *Isis* 74(1), 38–52.

[50] Descartes' relation to teleology is the matter of some debate; see especially (and references therein): Simmons, Alison J. (2001), "Sensible ends: Latent teleology in Descartes' account of sensation." *Journal of the History of Philosophy* 39.1 (2001): 49–75; Chapter 8 of Detlefsen, Karen (2013), *Descartes' Meditations: a critical guide*, Cambridge: Cambridge University Press; and Distelzweig, Peter (2015), "The Uses of Usus and the Function of Functio: Teleology and its Limits in Descartes' Physiology," *Journal of the History of Philosophy.* See also Manning, Gideon (2013), *Descartes' healthy machines and the human exception*, Leiden: Springer.

[51] Harvey 1651, Ex.30, 90–91.

being obvious from inspection: it is built into his observational language, in just the way that Longino points out is normal for bridging the semantic gap: the background assumption is built right into his conception of the phenomena.

Thus, if Harvey describes generation as wise, he must also explain this wisdom, and so this natural theological *cum* teleological background assumption affects his causal understanding of generation. Given this background assumption, it should be no surprise that God is, in fact, the first efficient cause: "For, in the construction of the chick the first efficient must use skill, foresight, wisdom, goodness and understanding far beyond the capacity of our rational souls."[52] Only God has the requisite foresight and art to explain the wondrous process of generation, and, in light of this assumption, any attempt to understand generation purely materially dishonors the Creator.[53]

1.4 Conclusions

Part of the large-scale shift that occurs across early modern natural philosophy in Europe involves the reconceptualization of matter. The foregoing account of background assumptions can pinpoint exactly those aspects to which Harvey, in particular, objected among this changing set of philosophical ideas. One can draw a direct line of influence (though not necessarily agreement) from Harvey's discussions of these issues to works of later materialist philosophers operating in England, such as Boyle in his *Disquisition on Final Causes* and John Ray in his *Wisdom of God,* and to later philosophers often classified as Vitalist, such as Margaret Cavendish.[54] Though they rejected Harvey's Aristotelian conception of explanation and science, they were in complete agreement with his characterization of generation, and the first two with his natural theology. Ray, for instance, cites his work on generation as empirically refuting Descartes' theory,[55] and Boyle (approvingly) notes the central importance of the idea of Divine Design in motivating Harvey's earlier investigation into the heart.[56] Further, as a number of commentators have pointed out since Jacques Roger's seminal work in the 1960s, over the course of the late seventeenth century, God becomes central to understanding life, matter, and for offering any sort of account or explanation of generation. As Dennis Des Chene has noted in his discussion of the Cartesian Regis:

---

[52] Harvey 1651, Ex.50, 144.

[53] See Goldberg 2013.

[54] See especially her 1666 *Observations upon Experimental Philosophy,* London, and her 1664 *Philosophical Letters,* London. In the latter she discusses Harvey's *De generatione.*

[55] E.g., Ray, John (1714), *The Wisdom of God as Manifested in the Works of Creation,* London, 45, 75.

[56] This is the famous passage where Boyle argues that the background assumption of Design, in particular the design of the valves, was key to Harvey's project. Boyle, Robert 1688, *A Disquisition about the Final Causes of Natural Things*, London: 157–158.

> Among the phenomena of life, generation offers...the greatest challenge to a science based on Cartesian principles. We know now that the means available to explanation of those phenomena were hopelessly inadequate to the task. Régis too believes that mechanism has its limits. But what lies beyond is not a revised conception of matter. What lies beyond is theology.[57]

This demonstrates the central importance of these background assumptions about God's relationship to nature, and helps contextualize how questions concerning Divine Design are fundamental to our accounts of debates over the proper method and theory of living phenomena in early modern Europe.

I want to end by discussing three conclusions that stem from this essay. The first concerns the context in which the concept of underdetermination is most often invoked: the realism debate. I think the example I've elaborated here demonstrates that realism, at least when it depends upon the concept of 'explanatory success,' doesn't really make any sense. This stems from the fact that empirical description and scientific explanation are not neatly separable: how Harvey thought about explanation affected his empirical descriptions, and his empirical work affected what he thought needed to be explained. What counts as success is highly contextual, and critically depends upon the language and conceptual system in which the theory is embedded. To return to the theme of this volume, this is a lesson that Peter taught—or better, told me—during my very first week of graduate school. In a long winded, often vulgar, hilarious, and irritatingly erudite rant, Peter long ago informed me of the historical poverty of the realism debate. It has only taken me a decade to get the message.

But, sticking with this theme, the Machamer way means philosophical concerns must stay central in any good history of science. Thus my second observation concerns historiographical categories like vitalism, materialism, mechanism, empiricism, rationalism, and so on. Much ink has been spilled on questions concerning the utility and applicability of these categories, asking, for instance, to what, exactly, does materialism refer?[58] or, what does the category of mechanism mean in context of living phenomena?[59] or, are non-actor's categories like empiricism and rationalism useful to understanding early modern natural philosophy?[60] and so on. In general, I think debates about these categories and concepts are important and useful as ways of understanding the large-scale structure and development of philosophy over time. Nevertheless, there are deep problems with their application to specific cases, problems that are only, perhaps, apparent to those trained in philosophy of science. It is here that background assumptions can be useful, for what is at stake in particular

[57] Des Chene 2003, 413.

[58] Wolfe 2014.

[59] Des Chene, Dennis (2005), "Mechanisms of life in the seventeenth century: Borelli, Perrault, Régis." *Studies in History and Philosophy of Science Part C: Studies in History and Philosophy of Biological and Biomedical Sciences* 36.2: 245–260.

[60] See the work of Peter Anstey and his students, e.g., Anstey, Peter R. (2005), "Experimental versus speculative natural philosophy," *Studies in History and Philosophy of Science* 19: 215–242; and, Anstey, Peter, and Alberto Vanzo (2012), "The origins of early modern experimental philosophy," *Intellectual History Review* 22.4: 499–518.

cases aren't historians' concepts and categories like vitalism or mechanism, but rather much smaller, more specific assumptions that affect how scientists *reason*, *experiment*, *interpret*, and, as I've think I've shown, *explain*. In the instance I've related, articulating Harvey's background assumptions helps us pinpoint his disagreement with 'those who philosophize these days,' in particular, how he thinks explanation should work, what he thinks needs to be explained, and what the larger (in this case, theological) stakes of explanation are. The point is not that the historiographical concepts are useless, but rather that we must philosophically interrogate the level at which we deploy them, not mistaking one level for another. Categories and concepts are high level, and are necessarily vague, whereas background assumptions are quite specific to individuals, perhaps even individual works.

Finally a small observation about scientific change. In the 1970s, Heinz Post coined the term 'Kuhn loss' to refer to Kuhn's statement that new paradigms inevitably lose some of the capabilities of the previous ones.[61] Something broadly similar can be seen in the history related here: the new materialistic philosophy could not explain the teleology so obvious to Harvey—obvious enough to affect his descriptions and observations—but what we see is not a new, mechanical way to explain these end-oriented features of nature. Instead, we see the elimination of teleological background assumptions, and, as a result, both a new way of conceiving nature as lacking natural ends in a way Harvey would understand, as well as a new way of describing nature, one that focused on motion and structure, and not upon organizing principles and operations. While this is a rather Kuhnian point, I don't mean to endorse the whole apparatus of paradigms. Indeed, there is an argument to be made that thinking in terms of underdetermination is preferable, perhaps even primary: background assumptions are smaller, discrete, detachable: they can be swapped out, combined, and mixed in ways that paradigms, as large, science-orienting worldviews, cannot. Indeed, it might be the case that paradigms were Kuhn's way of interpreting the fundamental problem of underdetermination at the level of scientific communities, and not at the level of individual theories and scientists, the level of the sorts of background assumptions I've discussed here.

As I mentioned above, it is not the case, as much recent historiography has revealed, that teleology goes away in scientific interpretations of nature. Instead it is the specific role that teleology plays in scientific explanation that changes, a shift from background assumptions like those of Harvey, where teleology is a, if not *the*, primary component of natural processes in need of explanation, to assumptions that merely identify teleology as an indicator of Divine Design. Harvey's second and third assumptions, which for him were deeply interconnected in a way typical for late Renaissance thinkers, came undone in the wake of new conceptions of philosophy and matter: what Harvey thought must be explained was no longer on the agenda. At least for some thinkers (though certainly not all), teleology had become

[61] Post, Heinz R. (1971), "Correspondence, invariance and heuristics: in praise of conservative induction," *Studies in History and Philosophy of Science Part A*, 2.3: 213–255. Kuhn says this in his 1962 [1996], *The Structure of Scientific Revolutions,* Chicago: University of Chicago Press, 169.

decoupled from the actual explanation of specific phenomena, and was viewed instead as important for characterizing the results of scientific observation in general, and for the characterization of God's relationship to the created world. Background assumptions, then, are useful ways for understanding and characterizing aspects of scientific change and debate. Some issues don't get resolved, aren't decided by accumulating observation and evidence, but are ignored and forgotten due to a shifting field of background assumptions.

References

Anstey, Peter R. 2005. Experimental Versus Speculative Natural Philosophy. *Studies in History and Philosophy of Science* 19: 215–242.

Anstey, Peter R., and Alberto Vanzo. 2012. The Origins of Early Modern Experimental Philosophy. *Intellectual History Review* 22(4): 499–518.

Boyle, Robert. 1688. *A Disquisition About the Final Causes of Natural Things*. London.

Cavendish, Margaret. 1664. *Philosophical Letters*. London.

———. 1666. *Observations Upon Experimental Philosophy*. London.

Cunningham, Andrew. 2002. The Pen and the Sword: Recovering the Disciplinary Identity of Physiology and Anatomy Before 1800 I: Old Physiology—The Pen. *Studies in History and Philosophy of Science Part C: Studies in the History and Philosophy of Biology and the Biomedical Sciences* 33(4): 631–665.

———. 2003. The Pen and the Sword: Recovering the Disciplinary Identity of Physiology and Anatomy Before 1800: Ii: Old Anatomy—The Sword. *Studies in History and Philosophy of Science Part C: Studies in History and Philosophy of Biological and Biomedical Sciences* 34(1): 51–76.

Dear, Peter. 2008. *The Intelligibility of Nature: How Science Makes Sense of the World*. Chicago: University of Chicago Press.

Des Chene, Dennis. 2003. Life After Descartes: Régis on Generation. *Perspectives on Science* 11(4): 410–420.

———. 2005. Mechanisms of Life in the Seventeenth Century: Borelli, Perrault, Régis. *Studies in History and Philosophy of Science Part C: Studies in History and Philosophy of Biological and Biomedical Sciences* 36(2): 245–260.

Detlefsen, Karen. 2013. *Descartes' Meditations: A Critical Guide*. Cambridge: Cambridge University Press.

Distelzweig, Peter. 2015. The Uses of Usus and the Function of Functio: Teleology and Its Limits in Descartes' Physiology. *Journal of the History of Philosophy* 53(3): 377–399.

———. 2016. Mechanics' and Mechanism in William Harvey's Anatomy: Varieties and Limits. In *Early Modern Medicine and Natural Philosophy*, ed. Peter Distelzweig, Benjamin Goldberg, and Evan Ragland. Dordrecht: Springer.

Doppelt, Gerald. 2007. The Value Ladenness of Scientific Knowledge. In *Value-Free Science? Ideals and Illusions*, ed. Harold Kincaid, John Dupré, and Alison Wylie. Oxford: Oxford University Press.

Easton, Patricia, and Melissa Gholamnejad. 2016. Louis De La Forge and the Development of Cartesian Medical Philosophy. In *Early Modern Medicine and Natural Philosophy*, ed. Peter Distelzweig, Benjamin Goldberg, and Evan Ragland. Dordrecht: Springer.

Feyerabend, P.K. 1965. On the 'Meaning' of Scientific Terms. *The Journal of Philosophy* 62(10): 266–274.

French, Roger. 1994. *William Harvey's Natural Philosophy*. Cambridge: Cambridge University Press.
Glanvill, Joseph. 1665. *Scepsis Scientifica; or, Confest Ignorance, the Way to Science: In an Essay on the Vanity of Dogmatizing and Confident Opinion*. London: E. Cotes for H. Eversden.
———. 1676. *Essays on Several Important Subjects in Philosophy and Religion*. London.
Goldberg, Benjamin. 2013. A Dark Business, Full of Shadows: Analogy and Theology in William Harvey. *Studies in History and Philosophy of Biological and Biomedical Sciences* 44: 419–432.
———. (Forthcoming). Anatomy as a Science of Teleology: The Case of William Harvey. In *Interpretations of Life in Heaven and Earth*, ed. Hiro Hirai. Dordrecht: Springer.
Gotthelf, Allan. 1987. Aristotle's Conception of Final Causality. In *Philosophical Issues in Aristotle's Biology*, ed. Allan Gotthelf and James Lennox. Cambridge: Cambridge University Press.
Gross, A.G. 1994a. Is a Rhetoric of Science Policy Possible? *Social Epistemology* 8: 273–280.
———. 1994b. The Roles of Rhetoric in the Public Understanding of Science. *Public Understanding of Science* 3: 3–23.
Harvey, William. 1616[1964]. *Prelectiones Anatomie Universalis*, ed. Gweneth Whitteridge. London: Royal College of Physicians.
———. 1628. *Exercitatio Anatomica De Motu Cordis et Sanguinis in Animalibus*. Frankfurt.
———. 1651. *Exercitationes De Generatione Animalium*. London.
Kincaid, Harold. 2007. Contextualist Morals and Science. In *Value-Free Science? Ideals and Illusions*, ed. Harold Kincaid, John Dupré, and Alison Wylie. Oxford: Oxford University Press.
Kuhn, Thomas. 1962 [1996]. *The Structure of Scientific Revolutions*. Chicago: University of Chicago Press.
Lakatos, Imre. 1970. *History of Science and Its Rational Reconstructions*. Proceedings of the Biennial Meeting of the Philosophy of Science Association. pp. 91–136.
Lennox, James. 1983. Robert Boyle's Defense of Teleological Inference in Experimental Science. *Isis* 74(1): 38–52.
———. 1985. Plato's Unnatural Teleology. In *Platonic Investigations*, ed. Dominic O'Meara. Washington, DC: Catholic University of America.
———. 2006. The Comparative Study of Animal Development: William Harvey's Aristotelianism. In *The Problem of Animal Generation in Early Modern Philosophy*, ed. J.E.H. Smith. Cambridge: Cambridge University Press.
Longino, Helen. 1990. *Science as Social Knowledge*. Princeton: Princeton University Press.
———. 2002. *The Fate of Knowledge*. Princeton: Princeton University Press.
Magnus, P.D. 2005a. Hormone Research as an Exemplar of Underdetermination. *Studies in History and Philosophy of Biological and Biomedical Sciences* 36: 559–567.
———. 2005b. Reckoning the Shape of Everything: Underdetermination and Cosmotopology. *British Journal for the Philosophy of Science* 56(3): 541–557.
Manning, Gideon. 2013. *Descartes' Healthy Machines and the Human Exception*, 237–262. Leiden: Springer.
McDonough, Jeffery K. 2011. The Heyday of Teleology and Early Modern Philosophy. *Midwest Studies in Philosophy* 35(1): 179–204.
———. 2016. Leibniz on Monadic Teleology and Optimal Form. In *Studia Leibnitiana Sonderhaft*, ed. Arnauld Pelletier. Stuttgart: Franz Steiner Verlag.
McIntyre, L.C. 2003. Taking Underdetermination Seriously. *SATS: Nordic Journal of Philosophy* 4(1): 59–72.
Okasha, Samir. 1997. Laudan and Leplin on Empirical Equivalence. *British Journal for the Philosophy of Science* 48: 251–256.
———. 2000. Underdetermination and the Strong Programme. *International Studies in the Philosophy of Science* 14(3): 283–297.
Post, Heinz R. 1971. Correspondence, Invariance and Heuristics: In Praise of Conservative Induction. *Studies in History and Philosophy of Science Part A* 2(3): 213–255.

Ray, John. 1714. The Wisdom of God as Manifested in the Works of Creation. London.
Roger, Jacques. 1963. *Les Sciences De La Vie Dans Le Pensée Françaese Du XVIIIe Siècle*. Paris: Armand Colin.
Roush, Sherrilyn. 2007. Constructive Empiricism and the Role of Social Values in Science. In *Value-Free Science? Ideals and Illusions*, ed. Harold Kincaid, John Dupré, and Alison Wylie. Oxford: Oxford University Press.
Ruphy, Stéphanie. 2006. "Empiricism All the Way Down": A Defense of the Value-Neutrality of Science in Response to Helen Longino's Contextual Empiricism. *Perspectives on Science* 14(2): 189–214.
Simmons, Alison J. 2001. Sensible Ends: Latent Teleology in Descartes' Account of Sensation. *Journal of the History of Philosophy* 39(1): 49–75.
von Haller, A. 1757–1766. Elementa Physiologiae Corporis Humani, vol.1: *Fibra, Vasa, Circuitus sanguinis, Cor*. Lausane.
Whitehead, A.N. 1953. *Science and the Modern World*. New York: The Free Press.
Wolfe, Charles T. 2014. Materialism. In *The Routledge Companion to 18th Century Philosophy*, ed. Aaron Garrett. New York: Routledge.

Chapter 2
Motion and Proportion in Simon Stevin's Mechanics

Maarten Van Dyck

Abstract This paper offers a reassesment of Simon Stevin's mechanics, by focusing on how Stevin tries to anchor his mathematical demonstrations in the behavior of material instruments. It is shown how his views on the relation between *spiegheling* (speculation) and *daet* (practice) are crucial to correctly understand his famous proof of the law of the inclined plane and his experimental test of the Aristotelian law of free fall. The distance separating *spiegheling* and *daet* is reproduced in that between instruments at rest and instruments in motion, because of Stevin's claim that impediments to motion are "inseperable accidents" of all moving objects.

2.1 Machines as Models of Intelligibility in the Archimedean Tradition

In his "Galileo's machines, his mathematics, and his experiments" (Machamer 1998), Peter Machamer proposed to call Galileo's use of the simple machines "models of intelligibility". These machines function as models of intelligibility by "directing attention to what is important in a problem and by exhibiting what relations exist among those important elements" (Machamer 1998, 72). As models they made it possible for Galileo to search in a directed way for the mathematical structures that could characterize phenomena of motion, such as the free fall or projection of bodies. The simple machines, among which the Archimedean balance occupied a privileged place, were ideally suited to this role: they are (1) material objects, (2) for which there existed successful mathematical treatments of the relations between their basic properties, and (3) that could be easily manipulated in a controlled way. They showed how to see ideal mathematical structures within concrete and material phenomena, a way of looking that was informed by the use of specific mathematical techniques (proportional geometry and Euclidean style proofs), representational conventions (allowing the coordination of the

M. Van Dyck (✉)
Department of Philosophy and Moral Science, Ghent University, Ghent, Belgium
e-mail: maarten.vandyck@ugent.be

M.P. Adams et al. (eds.), *Eppur si muove: Doing History and Philosophy of Science with Peter Machamer*, The Western Ontario Series in Philosophy of Science 81, DOI 10.1007/978-3-319-52768-0_2

phenomenon with the mathematical structure), and a practice in which the objects treated were actively manipulated and dealt with to achieve certain goals.

Machamer suggests that this focus on machines as models of intelligibility puts Galileo squarely within an Archimedean tradition (Machamer 1998, 56–57). As with the identification of any supposed 'tradition', it is clear that this claim must be qualified in a number of ways (see (Laird 1991; Hoyrup 1992; Bertoloni Meli 1992; Van Dyck 2006)), but there is no doubt that in general terms it is a useful construct and that Machamer's models of intelligibility capture a crucial aspect of this usefulness. Given the existence of such an Archimedean tradition, Simon Stevin was at least as much part of it as was Galileo.[1] Both were highly competent and innovative mathematicians who stressed the importance of actually constructing the machines described in theoretical terms. Yet as I will show, their way of seeing the ideal mathematical structure within concrete and material devices differed in a crucial respect. And this difference seriously circumscribed the realm of phenomena for which the simple machines could be used as models of intelligibility for Stevin. This is the main conclusion I want to draw: that depending on how one interprets the conditions that allow for the successful treatment of an instrument such as the Archimedean balance, one will hold very different views on the possibilities of extending it to different kinds of phenomena. Along the way I will try to come to a better understanding of two much celebrated aspects of Stevin's mechanics: his ingenious proof of the law of the inclined plane, and his experimental test of Aristotle's law of free fall; and of one often criticized aspect: his refusal to accord any weight to considerations based on so-called virtual diplacements.

2.2 Speculation and Practice in Stevin's Mechanics

When Stevin published his groundbreaking work on mechanics in 1586, he did so in one volume containing three separate treatises, *The Principles of the Art of Weighing*, *The Practice of Weighing*, and *The Principles of the Weight of Water*, which included a *Preamble to the Practice of the Weight of Water* and an *Appendix* to the three treatises.[2] "Practice" translates *daet*, which comprises a wide range of meanings, such as "effect", "work", and "construction". The books containing principles are examples of what Stevin calls *spiegheling*, the Dutch term he introduces

[1] Simon Stevin was born in Flanders in 1548 and died in the Dutch Republic in 1620, which makes him about one generation older than Galileo. The best general treatment (although outdated on a number of points) of Stevin remains Dijksterhuis' monograph from 1943, which was translated in an abridged version in 1970 (Dijksterhuis 1943; Dijksterhuis 1970).

[2] I use "mechanics" here as general name for the subject matter of the whole of Stevin's treatises. Stevin himself reserves the term "mechanical" for operations involving practical constructions, and stresses that the mathematical theory is rather one of the "free arts". All three treatises are included with English translation in Volume 1 of the modern edition of Stevin's *Principal Works* (Stevin 1955).

for *theoria*.[3] In a later text, he gives a concise description of the relation between *spiegheling* and *daet*:

> Speculation is an imaginary operation without natural matter … Practice is an operation which essentially takes place with natural matter. The conclusion of speculative propositions is perfect, but that of practical propositions is imperfect. … The property and the end of speculation is that it furnishes a sure foundation for the method of construction in practice, in which by closer and more painstaking care one may get as near to the perfection of the speculation as the purpose of the matter requires for the benefit of man. (Stevin 1961, 619)[4]

It is crucial that speculation should be oriented towards practical constructions, but Stevin also stresses the importance of clearly separating speculation and construction (Stevin 1961, 619). Speculation is not merely an instrument to achieve practical success, it should also render the practice intelligible by uncovering the causes behind its success. And to that end it needs to be allowed to operate within its own realm defined by appropriate idealizations.

In accordance with this general idea, Stevin introduces his *Principles of the Art of Weighing* with the statement that he will deal with "gravity, dissociated in thought from natural matter" (Stevin 1955, 95). This allows him to introduce a number of definitions characterizing the abstract properties of bodies' weight. The most important one is that of centre of gravity, taken over from Pappus via Commandino: "The centre of gravity is the point such that if a solid is conceived to be suspended from it, it remains at rest in any position given to it." (Stevin 1955, 99) This theoretical notion provides the basis for Stevin's mathematical treatment of the different properties of the equilibrium of bodies, starting from a proof of the law of the lever. On introducing the definition, Stevin highlights its speculative character, since it is based on an imaginary operation in thought. Accordingly, his methodological outlook implies that this speculative treatment must derive its legitimacy from the possibility of translating it back into practice. This is guaranteed in three ways by Stevin.[5]

Firstly, he notes that strictly (i.e., "mathematically") speaking all bodies will only have a unique centre of gravity if the forces of weight of different parts of a body are considered to be acting in parallel, perpendicular to the horizon considered as a plane, rather than converging in the centre of the world, as is actually the case; but he also claims that since the difference caused by this convergence is "imperceptible in all things weighed by us", we are justified in postulating them to be parallel

---

[3] The English terms that best translate Stevin's are "reflection" and "speculation": a spieghel is a mirror, and *spiegheling* can be both the image reflected in a mirror, and the act of mirroring. I'll translate it as "speculation".

[4] I altered the translation from the *Principal Works* at a number of places, as I have also done in further quotations. Page numbers are given to the English translations for ease of reference.

[5] It's interesting to remark that all three ways in which Stevin aligns his theoretical treatment with material practice are also to be found in the slightly earlier work on mechanics of the Italian mathematician Guidobaldo del Monte (see (Van Dyck 2006) for a detailed analysis on all three points – see also (Bertoloni Meli 2006, 26–32)). We'll come back to a further, related similarity between Stevin and Guidobaldo in Sect. 2.4.

(Stevin 1955, 113–115). The criterion for theoretical relevance is thus practical difference. Secondly, he is careful to prove that his mathematically demonstrated propositions also hold for bodies having irregular shapes, even if for simplicity he always considers bodies with regular shapes in his demonstrations. Significantly, this is proven twice in his treatise: a first time right after the first proposition (which introduces the law of the lever), and a second time in the very last proposition of the book, where it is argued that all previous propositions also apply to irregular bodies (Stevin 1955, 119, 219). In this way, this last proposition provides a natural bridge to the book on *The Practice of Weighing* that follows, where material bodies with irregular natural shapes are dealt with. Thirdly, and most importantly, he shows how the imaginary operation in the definition of centre of gravity can be translated into a practical operation. In his propositions 6, 7, and 8, Stevin deals with the cases of what we would call stable, indifferent and unstable equilibrium, where depending on the relative position of a body's centre of gravity and the fulcrum which supports it, the body will react differently when it is turned around its fulcrum and then released. He immediately notes that "if anyone should wish to see the contents of the preceding three propositions by some experience", it is easy to construct a small device illustrating the three cases, using nothing more than a wooden ruler with three holes in it and a needle acting as fulcrum (Stevin 1955, 143). The crucial case is that of indifferent equilibrium, which arises when a body's centre of gravity coincides with the position of the fulcrum: the body will remain in equilibrium whatever the position given to it. This possibility was denied by earlier writers such as Cardano, but it actually follows directly from the definition of the centre of gravity.[6] Its empirical possibility can thus be seen as a direct proof of the legitimacy of the latter definition. It is then not surprising that proposition 2 of *The Practice of Weighing* deals in considerable detail with the construction of a real life balance that exemplifies indifferent equilibrium, and which Stevin designates as "a most perfect balance" (Stevin 1955, 303).

The superiority of Stevin's treatment of mechanics over all earlier and contemporaneous authors lies in the way in which considerations based on the bodies' centre of gravity are embedded in a general framework that allows the analysis of the statical effects of forces applied in arbitrary directions.[7] The crucial step towards

[6]As we will see below, Cardano is explicitly cited at two crucial places in Stevin's *Appendix*. Apart from a passing reference to Jacques Besson and the curious case of Jean Taisnier (see Sect. 2.5), Cardano and Commandino are the only contemporary authors in mechanics who are cited by Stevin. Whereas the latter provided an exemplar, the former served as a foil. In the same work cited by Stevin in the Appendix, the *Opus Novum de Proportionibus*... of 1570, Cardano had not only given an incorrect law of the inclined plane, he had also denied indifferent equilibrium and had claimed that one should consider the lines of weight as converging in the centre of the earth (Cardano 1570). In his *Appendix*, Stevin indicated that he wanted to refrain from explicitly engaging with other authors in his main text, but it is fair to assume that in the passages just referred to in the text he has Cardano in view as one of the authors who "practised many false propositions" (Stevin 1955, 509).

[7]Dijksterhuis provides a succinct overview in (Dijksterhuis 1970, 54–57). The explanatory scheme in Guidobaldo del Monte's *Liber Mechanicorum*, the only contemporaneous book that can rival

the general rule for the composition of forces that is implicit in these analyses, is provided by the law of the inclined plane, since it relates the oblique force parallel to the inclined plane to the vertical weight and its component normal to the inclined plane (in a typical further move the plane itself will be abstracted, or "thought away", after having played its initial role). Well aware of the novelty of his treatment, and consistent with his general methodological orientation, Stevin again complements his mathematical demonstrations with a description of how to build a simple device that allows us "to see, examine, and understand through visual experience the truth of the propositions on the proportionality of such weights, described mathematically in the first book, in order that we may have all the more confidence in that which we wish to effect therewith in practice for the benefit of mankind." (Stevin 1955, 321) Indeed, it seems clear that one of the main aims that Stevin set himself with the publication of his books on mechanics was to teach his readers how to see ideal mathematical structures within concrete and material phenomena, a way of seeing that both implied a clear separation of speculation from practice and a constructive anchoring of the former in the latter.[8]

Let us now turn towards Stevin's mathematical proof of the inclined plane law, no doubt the most celebrated part of his mechanics.

2.3 Proving the Law of the Inclined Plane

2.3.1 *Different Forms of Perpetual Motion*

To understand a number of interpretative problems that have occupied modern commentators, it is useful to start with laying out some basic analytical distinctions concerning perpetual motion as understood from a present-day perspective. A good starting point is the following classification proposed by Alan Gabbey (Gabbey 1985, 42–43):

> PM1. "A system or device that maintains indefinitely the motion it already has, in the absence of dissipative influences. Examples: a single body moving "inertially" through empty Newtonian space; ..."
>
> PM2. "A device that moves perpetually with the same motion or the same repeated sequence of movements, overcoming dissipative influences alone, or in addition doing useful work of some kind. The device receives no power from sources other than the *inertia* (*impetus*, in pre-Newtonian terminology), or the *weight* and inertia, of its constituent

Stevin's in terms of rigor, breaks down exactly with forces that are not applied in the direction of weight (Van Dyck 2006, 394–96; Renn and Damerow 2012).

[8] Stevin himself often comments on pedagogical issues, and presents the most important collection of his works (the *Mathematical Memoirs*) as the set of courses he used with Prince Maurice. Approaching his work from this angle also allows us to place Stevin's work squarely in its historical context, more specifically that of late humanism at Leiden, which was characterized by an intense pedagogical focus on the proper interrelation of theory and practice (van Bunge 2001, chapter 1).

bodies. Examples: Bhaskara's "mercury wheels" (c. 1150), Villard de Honnecourt's "mallet wheel" (1235); …"

From our contemporary perspective, the denial of the possibility of perpetual motion comes down to the denial of PM2. According to Gabbey, in the seventeenth century these devices "were held to be impossible, both in theory and in practice, by those who were alive to certain implications of Peripatetic natural philosophy, and by those who were au fait with the principles of mechanics." (Gabbey 1985, 45) The status of PM1 was less clear-cut, though, as is clear from its close association with "inertial" motion.

In what follows I will bring out a crucial ambiguity in Stevin's denial of the possibility of perpetual motion in his famous *clootcrans* (wreath of spheres) proof: whether he intends to deny PM1 or PM2. In Sect. 2.4, I will then argue that this ambiguity is not accidental but closely linked to the sharp distinction Stevin maintains between the speculative and the practical part of mechanics.

2.3.2 The **Clootcrans** *Proof*

The law of the inclined plane states that two bodies lying on the upper sides of horizontally placed triangular prism will be in equilibrium if their absolute weights stand in the same proportion as the lengths of the respective sides on which they are placed. To prove the law, Stevin considers a wreath of spheres that is hung around the prism. The kernel of his proof consists of the claim that if the stated conditions hold for the weights of the combined spheres on the respective upper sides, there must be equilibrium – because otherwise the wreath would "spontaneously" perform a perpetual motion, which according to Stevin is "false" (Stevin 1955, 178) (see Fig. 2.1).[9]

In his notes to the modern edition of Stevin's work, E. J. Dijksterhuis remarks: "The conviction that a perpetual motion is impossible in physical reality is not a sufficient ground for qualifying it as absurd in the ideal sphere of rational mechanics,

[9] The translation in the *Principal Works* opts for "automatically" and "absurd" respectively (Stevin 1955, 179). Stevin's Dutch literarily states that the spheres perform a perpetual motion "out of themselves" ("uyt haer selven"). The 1605 Latin translation by Willebrord Snell has "ex sese" and "falsum" respectively (Stevin 1605, 35). Girard's 1634 French edition states "ce mouvement n'auroit aucune fin, ce qui est absurde", leaving out all explicit reference to the qualification that the motion is self-caused (Stevin 1634, 448). As will become clear in Sect. 2.4.3, this qualification is crucial for a good understanding of the grounds of Stevin's argument. As for "absurd", in this context that is a perfectly acceptable translation of Stevin's "*valsch*", but we will see in Sect. 2.4 that the distinction between "truthful" and "false" plays a central role in Stevin's thinking about acceptable idealizations within the science of mechanics. As I will also argue that this distinction is crucially linked to the ambiguity I will uncover in Stevin's appeal to this "absurdity" or "falsehood", it is preferable to keep the terminological link explicit. (In his hydrostatics, Stevin calls the possibility of perpetual motion "*ongeschickt*" (which at other places in his work he uses both to mean "absurd" and to literally mean "irregular") (Stevin 1955, 400).

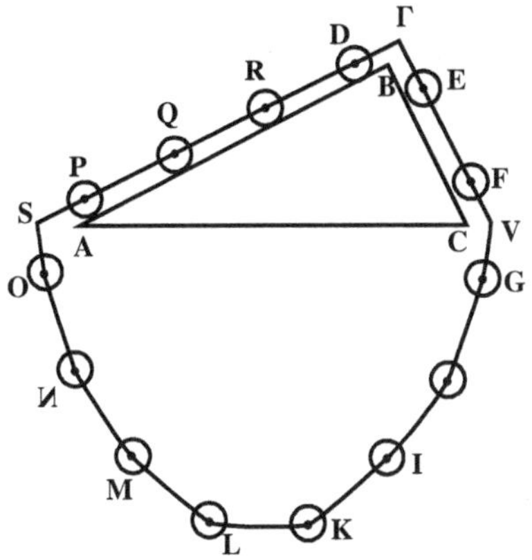

Fig. 2.1 Imagine that the conditions stated in the law of the inclined plane hold for the combined weights *PQRD* and *EF* respectively, but that there would be no equilibrium. The geometry of the situation implies that the resulting motion of the clootcrans reproduces the original configuration, and that there would again be disequilibrium, such that again motion would follow

where friction and resistance of the air are absent." (Stevin 1955, 179)[10] On this interpretation, Stevin makes the mistake of denying the possibility of PM1, basing his proof on what is at most an empirical generalization of observed phenomena (with friction and resistance present) that is insufficient to carry the weight of a rational proof. Alan Gabbey gives a related criticism when he states that "curiously, it has not often been noted that Stevin's demonstration is invalid, and that its invalidity derives from the fact that within the terms he himself would have accepted, which were those of pre-Newtonian mechanics, such a perpetual motion would not be absurd at all… The consequent of Stevin's counterfactual supposition is an example of constant motion of an idealized closed cycle of bodies under a constant force, which was a straightforward situation in pre-Newtonian physics." (Gabbey 1985, 74)

Let me begin with noticing that it is not so evident that Stevin's hypothetical situation could find a place within a pre-Newtonian context. The only more or less "straightforward" cases of perpetual (and not just constant) motions caused by a constant force were those of heavenly bodies around the centre of the world, where the moving force is certainly not provided by the weight of the bodies. But if we ignore this complication, Gabbey seems to have a good point when claiming that Stevin "would have accepted" that a constant force causes a constant motion. Evidence is not hard to find: Stevin explicitly states that the motion of the wreath "will continue for ever, because the cause is always the same" (Stevin 1955, 179). This seems to imply that the continuously operating force doesn't result in a continuous acceleration, but rather in a uniform motion, which would again mean that Stevin's proof rests on an unwarranted denial of the possibility of PM1.

Yet, before we accept this analysis, we need to ask ourselves what are the conditions under which the motion of the wreath is supposed to take place. Both

[10] In (Dijksterhuis 1943, 121–23), Dijksterhuis expands on this criticism, and identifies some further logical gaps in the proof as presented by Stevin (the truncated discussion in (Dijksterhuis 1970, 54) only retains the criticism that the proof seems to rest on an invalid denial of PM1).

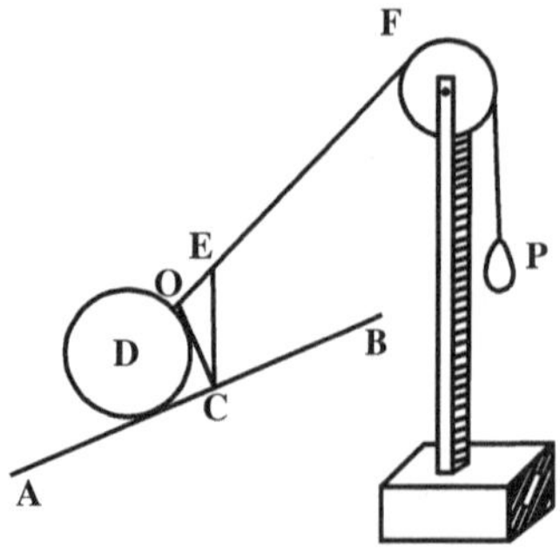

Fig. 2.2 The law of the inclined plane implies that there is equilibrium if body *P* has the same ratio to *D* as length *EO* to *CE*, but if the plane is horizontal points *E* and *O* coincide

Dijksterhuis and Gabbey assume that there is no friction and air resistance to be overcome; and in so far as that they pay attention to the question, all other commentators seem to follow them in this.[11] It is striking, though, that Stevin nowhere explicitly states this to be the case. And if there is friction, then work would need to be continually performed to guarantee the continuing motion, and we would be presented with a case of PM2.

The least that we can conclude is that Stevin's own formulation leaves open a most crucial aspect, and that every univocal interpretation of the physical import of his argument depends on (possibly disputable) interpretation. At the same time Stevin's argument is probably completely convincing to most readers. It is exactly because of this kind of tension that Ernst Mach concludes in his analysis of the argument that its persuasiveness must find its origin in pre-theoretical, "instinctive" knowledge (Mach 1919, 26). According to Mach, we all feel that the situation described is impossible, even if we cannot explicate the grounds thereof in full clarity, as Stevin could not either.[12] In Sect. 2.4.3, I will try to throw some further light on the precise nature of the intuition that Stevin was appealing to.

2.4 Rest and Motion in Stevin's Mechanics

2.4.1 Motion on a Horizontal Plane

There is one passage where Stevin comes close to a conclusion that could have led him to clarify the ambiguity we just uncovered. At the end of the sixth corollary to the proposition in which he proves his law of the inclined plane, he asks what the proportionality expressed by it implies for a body that lies on a perfectly horizontal plane. In this case the mathematical ratio between the two geometrical magnitudes in the law is indeterminate (since the height has zero length) (see Fig. 2.2).

[11] Dijksterhuis has already been quoted to this effect. Gabbey also states that Stevin assumes that the spheres are "frictionless" (Gabbey 1985, 74).

[12] Notwithstanding his criticism of Stevin's formulation, Dijksterhuis comes to a very similar appreciation (Dijksterhuis 1943, 122–23; Dijksterhuis 1970, 54).

According to Stevin this implies that, whatever the ratio of the weights of the two objects, they will never have equal apparent weight. Stevin interprets this as follows:

> by which it is to be understood that a gravity taking the place of *P*, however small it may be, cannot be of equally apparent weight to the solid *D*, but will pull it along (mathematically speaking), however heavy it may be. From this it follows that all gravities pulled along parallel to the horizon, such as ships in the water, wagons along the level land, etc. , to be moved do not require the force of a fly beyond that which is caused by the surrounding obstacles, viz. water, air, contact of the axles with the bearings, contact of the wheels with the road, and the like. (Stevin 1955, 187)

Stevin was not the first to come to the conclusion that a body can be moved along the horizontal by the smallest possible force (among his predecessors we also find Cardano, with whose work Stevin was certainly familiar).[13] In the hands of Galileo, who came to this idea in a similar way, it would acquire much greater mechanical and cosmological relevance: he went on to claim that since the conclusion implies that a body has no intrinsic tendency to rest on the horizontal, in the absence of all impediments it would also conserve the motion initially imparted to it. To put it differently, he would start using the idea in a way akin to an inertial principle (and this both in demonstrating the parabolic trajectory of projectiles, and in answering purported refutations of the possible motion of the earth). No doubt, this is one of the most spectacular instances of Galileo's use of models of intelligibility that stem from the science of machines. And crucially, Stevin doesn't take this further step. He never considers what would happen with the body in the absence of all impediments, so neither does he reflect on what his "mathematical" conclusion implies for the relation between force and the continuation of motion.[14] As will become clear, this silence is due to the fact that Stevin considers the question what a moving body would do in the absence of all impediments as devoid of sense.

2.4.2 **Spiegheling** *and* **Daet,** ***Rest and Motion***

The short passage on motion on the horizontal plane is one of the few places in *The Art of Weighing* where Stevin speaks of motion. The treatise is completely devoted to the determination of conditions for equilibrium, whereas machines that have to put bodies in motion belong to the domain of *The Practice of Weighing*. It is probably not accidental that just before this passage, Stevin introduces an explicit (and given the structure of the work, somewhat surprising) reference to *The Practice of Weighing*, by claiming that all demonstrated corollaries to the law of the inclined

---

[13] See (Festa and Roux 2006) for different examples, both in antiquity (Heron) and in the sixteenth century, and further analysis; surprisingly, Stevin is not mentioned.

[14] This also shows in the fact that when he replies to the same anti-Copernican arguments as Galileo, his discussion is not connected to these mechanical considerations at all (Stevin 1961, 125–27).

plane are "true not only of spheres, but also of other solids sliding or rolling on points or lines … (a subject with which we will deal more properly in the Practice of Weighing)." (Stevin 1955, 185)

The Practice of Weighing itself opens with the following warning:

> Because in several propositions of the Practice of Weighing the motions of bodies will be dealt with, I thought it advisable, before coming to the matter, to explain something of it to the reader. To wit, that the Art of Weighing only teaches us to bring the moving body into equality of apparent weight to the body to be moved. As to the additional weight or the force which the moving body requires in order to set in motion the body to be moved (which weight or force has to overcome the impediments of the body to be moved, which is an inseparable attribute of every body to be moved), the Art of Weighing does not teach us to find that weight or force mathematically; the cause of this is that the one moved body and its impediment are not proportional to the other moved body and its impediment. (Stevin 1955, 297)

But this need not be a cause of concern, he explains:

> Note also that this knowledge of equality of apparent weight is sufficient for the purpose, for if the same weight lies in either pan of the balance, as we then know (though the balance also has its impediment to motion) that little force is required to move the pans, thus it is also in all other cases. (Stevin 1955, 299)

The law of the lever and the law of the inclined plane suffice to give mathematical determinations of the conditions of equilibrium in a wide variety of circumstances. Stevin moreover claims that, just as in the case of the object on the horizontal plane, the smallest imaginable weight would be enough to put a balance in indifferent equilibrium into motion, "mathematically speaking" (Stevin 1955, 307). But in practice, we always need to add as much weight as is needed to overcome the ever present impediments. And the latter can only be determined empirically and on a case per case basis.

A year before his treatises on mechanics, Stevin had already published a Dutch *Dialectike or the Art of Demonstrations*, in which he gave an explicit (and completely traditional) definition of an "inseparable accident":

> INSEPARABLE ACCIDENT, *is that which cannot be taken away from its subject without the demise of the thing*, as life from an animal, because if that is gone there is no longer an animal, but a dead body; or also the plane from a solid, the form of matter … (Stevin 1585, 21)

In the very first sentence of his *The Art of Weighing* Stevin had called number, magnitude and weight inseperable accidents of all existing things, each giving rise to a specific mathematical science (respectively arithmetic, geometry and mechanics or the art of weighing) (Stevin 1955, 55). By calling the impediments to motion an inseparable accident of bodies to be moved at the beginning of *The Practice of Weighing*, Stevin excludes the possibility of coming to a legitimate mathematical science of motion that would abstract away these impediments.

This throws further light on the relation between *The Art of Weighing* and *The Practice of Weighing*, between *spiegheling* and *daet*. As already quoted in Sect. 2.2, "speculation is an imaginary operation without natural matter", whereas "practice is an operation which essentially takes place with natural matter" (Stevin 1961, 619);

and in *The Art of Weighing* we consider bodies to which we ascribe the inseperable properties of magnitude and weight, but "dissociated in thought from natural matter" (Stevin 1955, 95). Under these counterfactual circumstances, we can give exact determinations of the conditions under which there is equilibrium. These results remain valid for material objects: if we add the "natural matter" back, this will certainly not change the fact that bodies for which these conditions hold remain in equilibrium. Stevin summarizes his dependence on counterfactual circumstances as follows: "The false is admitted in order that the truthful may be learned therefrom." (Stevin 1955, 227) For moving bodies, we can no longer do this: the addition of "natural matter" does render all proposed exact proportions on motion false since motion inevitably brings with it impediments. Ignoring this point condemns one to "holding the false for truthful" (Stevin 1955, 299).

This makes clear why Stevin is not interested in what would happen with a body on a horizontal plane in the case in which all impediments are considered to be absent. This is a purely hypothetical situation that can be no "mirror" of any possible practical operation, whereas all speculation must always be oriented towards "common practical weighings" (Stevin 1955, 521). Hypotheses on the behavior of moving objects in the absence of all impediments would belong to what he calls (in another context) the arsenal of "mirror-fighters, who violently fight against their shadow without touching it" (Stevin 1964, 44). The bridge between the mathematical and the material can only be built for equilibrium situations. The "ideal sphere of rational mechanics" that Dijksterhuis referred to (see Sect. 2.3.2) unfolds itself in a way fundamentally different than it will with later authors.[15]

An example that we already came across is Galileo's use of the law of the inclined plane as a model of intelligibility from which he extrapolates his "proto-inertial" principle. This move implies that Galileo does ignore the sharp boundary that Stevin sees between machines in equilibrium and bodies in motion. He considers the frictionless situation as nothing more than the limit situation of all observed instances of moving bodies, without any change in the nature of the phenomenon under consideration. As a practicing engineer, Stevin was perfectly capable of seeing the importance and possibility of increasingly reducing friction by greasing and polishing the materials, but as he had already stated in his *Dialectike*, "separable and inseparable accident do allow some mean, such as the heat of fire and the like, which can be taken away in large part from its subject, yet without being completely separable" (Stevin 1585, 21). The practical possibility of reducing their effect does not turn the impediments into "separable accidents", as they have become for Galileo. It is only with the latter move that the behavior that results when we "think

[15] One author who held the very same position as Stevin on this crucial point was, again, Guidobaldo del Monte (see (Van Dyck 2006; Laird 2013)). Stevin almost certainly knew Guidobaldo's treatise on perspective from 1600 when he published his own in 1605 (see (Andersen 1990)), but there are no indications he was familiar with the Italian's treatise on mechanics in 1586. It cannot be excluded, but the absence of any attention to pulley systems makes it rather unlikely: pulleys take up a large part of Guidobaldo's treatise, but Stevin only added an analysis in a supplement in 1605, reportedly after prince Maurice had seen a treatment thereof in Buonaiuto Lorini's *Delle Fortificationi*, a work dependent on Guidobaldo for its treatment of the simple machines.

them away" can retain its validity even in the absence of a direct empirical confirmation.[16]

2.4.3 *The* Clootcrans *Proof Reconsidered*

Let us now return to the *clootcrans* proof. Given its position in a treatise in which Stevin has "thought away" all natural matter, friction should not be present in the hypothetical motion. But given his conviction that friction is an inseparable accident for all moving bodies, it is not clear how the consideration of motion without friction can teach us anything physically relevant; and the characterization of the perpetual motion as "false" is certainly supposed to depend on a physical intuition![17]

It is clear that the appeal to perpetual motion occupies an exceptional and at first sight problematic place in Stevin's speculative treatise. Maybe it is exactly because it cannot take place that it can be introduced, and that Stevin at the same time can stay silent on its precise characteristics. After all it is excluded rather than analyzed.

But if this is true, then we should not be looking at considerations on the general relationship between forces and motion if we want to better understand Stevin's appeal to the impossibility of a perpetual motion. Its exclusion ought to be independent from the question whether there are dissipative influences to be overcome or not, whether work can be performed or not. Stevin always had an eye on practice when formulating his theories, but "work" as an intuitive notion was not a mathematizable aspect of the operations. It belongs to the realm of doing rather than knowing for him, and in no way it can provide the necessary foundations for the speculative part of mechanics.

So how should we understand Stevin's exclusion of the perpetual motion of the *clootcrans*? I think the most important clue is his claim that the motion of the wreath of spheres is not only perpetual, but that it is not started by an external force either but rather "by themselves". The only situation in which inanimate material objects start to move spontaneously is when they are in a state of disequilibrium, like a balance where the center of gravity is situated outside the vertical line under or above the fulcrum. But in these cases the objects' motions always stop when equilibrium is reached, i.e., when the centre of gravity has taken its preferred position. This is exactly the physical kernel of Stevin's propositions 6, 7, 8 that play such a crucial role in grounding the mathematical framework characterizing the equilibrium of

[16] This obviously raises the further question what lies behind this change in perspective, a question that cannot be addressed here. Let me just stress that it is easy to state that in the absence of all impediments, a body would have a certain behavior – but quite a bit more challenging to explain why we should think that this statement teaches us anything significant concerning the behavior of bodies in empirical circumstances!

[17] This is why I chose not to render Stevin's "*valsch*" as "absurd" (see footnote 9); the former more than the latter stresses its basis in physical considerations, as his use of "false" for empirically invalid statements (see the examples in Sect. 2.4.2) makes clear.

balances (see Sect. 2.2). The *clootcrans* proof extends this insight in a way that allows it to ground the case of equilibrium of bodies on an inclined plane. The case of the self-moving wreath of spheres would be one of a perpetually frustrated disequilibrium without the possibility of a corresponding equilibrium, and this is what is excluded.

This way of understanding Stevin's intuition not only has the virtue of not relying on the moot question whether friction has to be overcome or not, it also forges a much closer connection between his treatment of the balance and the inclined plane. The mathematical law of the lever is explicitly grounded in the fact that all material bodies have a centre of gravity – a fact that shows itself in how these bodies always search for a preferred position of equilibrium and don't leave this once it is reached. We can now see that the mathematical law of the inclined plane is implicitly grounded in the fact that a body will only start to move spontaneously if by doing so it can reach equilibrium – a claim that later would be made explicit in Torricelli's pinciple (bodies that are connected together will only start to move if they can lower their common centre of gravity). Stevin's intuition rests on the same experience of material bodies that was already conceptually encoded in the ascription of a centre of gravity to each body.

2.5 Testing the Aristotelian Law of Fall?

In the second chapter of the *Appendix* to his mechanical treatises, Stevin offers "a more appropriate proof" of what we just saw to be the crucial claim in the preface to *The Art of Weighing*, that "bodies in motion are not proportional to their impediments" (Stevin 1955, 509). Surprisingly he does this by empirically refuting authors who held that there was a proportionality to be found within the motion of falling bodies. The authors he cites are Aristotle, Johannes Taisnier,[18] and Cardano. He begins by claiming that Aristotle and his followers hold "that if two similar bodies fall through the air, as the gravity of the one is to that of the other, so is the time of passage of the latter to the former, i.e., as gravity is to gravity, so is impediment to impediment" (Stevin 1955, 509–11). In the modern edition to Stevin's work the last part of the statement is annotated: "It is by no means clear, how this inference is drawn"; but it actually shows that Stevin was a careful reader of Aristotle.[19] In the eight chapter of the fourth book of the *Physics*, Aristotle explains that there are differences in the fall of bodies caused by the medium (for a body falling through different media) and differences that are due to the weight of bodies (for two bodies

[18] This author had published a treatise in which he had actually presented the text of Giovanni Battista Benedetti's *Demonstratio proportionum motuum localium contra Aristotilem et omnes philosophes* from 1554. Stevin was unaware of this plagiarism and ascribed the ideas to Taisnier.

[19] In his *Appendix*, all references to Aristotle are directly taken over from Taisnier/Benedetti; but in the *Dialectike*, Stevin's related treatment of the problem of fall shows that he was indeed familiar with the relevant parts of Aristotle's *Physics* (Stevin 1585, 144–48).

falling through the same medium). But in explaining the latter factor it turns out that it actually causes proportional differences in speeds because the heavier body divides the medium through which they are moving "faster by its force" (Aristotle 2001, 285). It is indeed sensible to read this with Stevin as claiming that gravities are (inversely) proportional to impediments (the heavier can move faster because it is less impeded). Stevin then claims that he can show by "*daetlicke ervaring*" (a highly significant turn of phrase that is probably best translated as "constructed experience"), that this is false. He famously relates:

> Let us take (as the very learned Mr. Jan Cornets de Groot, most industrious investigator of the secrets of Nature, and myself have done) two spheres of lead, the one ten times larger and heavier than the other, and drop them together from a height of 30 feet on to a board or something on which they give a perceptible sound. Then it will be found that the lighter will not be ten times longer on its way than the heavier, but that they fall together on to the board so simultaneously that their two sounds seem to be one and the same rap. (Stevin 1955, 511)

At this point it is of course tempting to read Stevin as a direct precursor to Galileo, as is often done, but this is only possible if one ignores everything that follows. Rather than concluding that this experience proves that bodies of the same matter fall with the same speeds (a position that Stevin ascribes to Taisnier and Cardano), he immediately goes on to add a further experiment to also disprove this conclusion!

> Take a small, single, short hair of cotton and a packet of the same, tightly tied together, weighing one pound and of similar form to the hair; drop these together from a height of five or six feet, and experience will show that the hair (in spite of the fact that its material is much more compact than that of the packet, in which there is much empty space of air) is at least twenty-five times longer on its way than the packet. (Stevin 1955, 511)

Stevin offers an explanation for the proportionally different effect that a medium would have on bodies of the same shape and material but of different volume, by noting that the surfaces (which provide a measure for the amount of friction felt) grow as squares, whereas the volumes (which provide a measure for the weight) grow as cubes. He does not explain why the effect is so much more noticeable for cotton than for lead, though. His general conclusion suggests strongly that he would rather see this as an illustration of the fact that the impediments will always affect bodies of different material differently:

> But since we have thus proved that there is no proportionality between bodies in motion and their impediments in the most obvious examples, where there is only simple friction of the surfaces against the air or against the water, there will a fortiori be no proportionality in less obvious examples of several materials, such as tools of wood, iron, and the like, for the former is oiled, the latter greased; the one can swell in moist weather, the other rust; all of which things (I omit many others) lighten or weight the motions of the tools. (Stevin 1955, 513)

This diversity of material properties provides the main reason why we cannot "mirror" concrete phenomena of motion in the language of mathematical proportions. Rather than being aimed at confirming particular laws of fall, Stevin's experiments are thus meant to illustrate the mathematically irreducible diversity of the material world. When we are dealing with motion, there is at most an illusory "appearance of proportionality" (Stevin 1955, 513).

2.6 Divine Proportion

A year before the publication of his mechanical treatises, Stevin had already included a discussion on the impossibility of establishing a proportionality characterizing freely falling bodies in a short dialogue at the end of his *Dialectike* (Stevin 1585, 145–48). In that context, the discussion directly leads up to one of his favorite themes: the superiority of the Dutch language. Indeed, throughout all of Stevin's writings, his favorite and most significant example illustrating this superiority is *everedenheyt*, his Dutch term for *proportio*.[20] As he explains, the Dutch term hardly needs a definition as it expresses its content directly: "Everedenheyt is de Evenheyt der Redenen"; in contrast to e.g. its imperspicuous Latin rendering: "Proportio est Rationum similitudo" (Stevin 1585, 156). In Stevin's view, this non-transparent character explains why people have such a hard time applying the concept correctly to empirical situations, as illustrated by Aristole's and Cardano's erroneous statements on falling bodies. This helps understand the exclamation with which he ends his discussion on free fall: "Oh Divine Proportion, what Gods you could make of us, since you reveal so little of your hidden features to so few people!" (Stevin 1585, 148) Ten pages later it is confirmed that the "few people" he refers to are actually the Dutch, since he now explicitly states that they can become "earthly Gods" because their superior language will allow them to achieve a "knowledge of causes" about many of nature's "hidden features" (Stevin 1585, 163–64). Stevin effortlessly slides from proportionality as a mathematical concept to causal terminology, a move no doubt facilitated by the flexibility of the Dutch term "reden": Stevin uses this term not only to refer to a mathematical ratio, but also to reason or argument, and to a cause. A proportionality is simultaneously a similarity (or equality) of ratio's, and a causal relation that can be uncovered by mathematical argument.

This brings us to one last (in)famous aspect of Stevin's mechanics. In the first chapter of his *Appendix*, Stevin criticizes so-called 'Aristotelian' attempts to prove the law of the lever by appealing to the motion that the bodies on the arms of the balance would have if put in motion (Stevin 1955, 509).[21] Stevin laconically points out that since bodies in equilibrium do not move, this postulated motion cannot be

[20] In Stevin's work on musical theory, never published during his lifetime, he speaks about the "infinite power" inherent in the Dutch word for proportionality (Stevin 1966, 427).

[21] Again, Cardano is one of the authors who believed in the validity of such proof (Cardano 1570, 34). This is not explicitly mentioned by Stevin, although he does refer to Cardano's erroneous statement of the law of the inclined plane in the same chapter of the *Appendix*. In the second chapter, in which he disproves the possibility of having laws of motion, he does single out Cardano for his faulty reliance on the "appearance of proportionality" in motion, which led him to errors "in a great many different propositions", which can be plausibly read as also referring to this proof of the law of the lever (Stevin 1955, 513).

the cause of the equality of their apparent weight. This is especially interesting if we want to understand what he did, in contrast, consider to be legitimate causal explanations. Stevin refers the reader who wants to come to a "knowledge of the causes" to the very first proposition of *The Art of Weighing*, where he introduced a variant of the original Archimedean proof of the law of the lever. This proof is essentially a way of transforming different cases of equilibrium into each other, showing the fundamental equivalence between the symmetrical case of bodies of equal weight at equal distances and the case of unequal bodies at distances in inverse proportion to the weights. These transformations exploit geometrical relationships between the bodies and certain elementary properties of centers of gravity.[22] The end result is that the unequal bodies on the balance are shown to be in equilibrium *because* the proportionality expressed by the law of the lever implies that they are similar in a crucial respect to bodies that are necessarily in equilibrium (equal bodies at equal distances). The fact that it uncovers this similarity underlying the proportionality is thus what gives the demonstration its causal content.

Proportionality is the beginning and end of Stevin's mathematical theories. Where there is proportionality, there is the possibility of an almost divine mathematical knowledge of causes. Where there is none, we are left with the possibility of human practice that can make the diversity of material circumstances manageable according to different degrees of success. *Seeing* mathematical structure in concrete material phenomena is always circumscribed: we can only mirror part of nature – the remaining part we have to *handle*.

References

Andersen, Kirsti. 1990. Stevin's Theory of Perspective: The Origins of a Dutch Academic Approach to Perspective. *Tractrix* 2: 25–62.

Aristotle. 2001. *The Basic Works*, ed. Richard McKeon. New York: The Modern Library.

Bertoloni Meli, Domenico. 1992. Guidobaldo Dal Monte and the Archimedean Revival. *Nuncius* 7(1): 3–34. doi:10.1163/182539192X00019.

———. 2006. *Thinking with Objects: The Transformation of Mechanics in the Seventeenth Century*. Baltimore: Johns Hopkins University Press.

Cardano, Hieronymus. 1570. *Opus Novum de Proportionibus Numerorum, Motuum, Ponderum, Sonorum ....* Basel: Henricus Petri.

Dijksterhuis, E.J. 1943. *Simon Stevin*. Den Haag: Martinus Nijhoff.

———. 1970. *Simon Stevin: Science in the Netherlands around 1600*. Den Haag: Martinus Nijhoff.

Festa, Egidio, and Sophie Roux. 2006. La Moindre Petite Force Suffit À Mouvoir Un Corps Sur L'horizontal. L'émergence D'un Principe Mécanique et Son Devenir Cosmologique. *Galilaeana* 3: 123–147.

Gabbey, Alan. 1985. The Mechanical Philosophy and Its Problems: Mechanical Explanations, Impenetrability, and Perpetual Motion. In *Change and Progress in Modern Science*, ed. Joseph

[22] Guidobaldo del Monte's detailed commentary of the slightly different original Archimedean proof is analyzed in (Van Dyck 2013).

C. Pitt, 9–84. Dordrecht: Reidel. http://link.springer.com/chapter/10.1007/978-94-009-6525-6_2.

Hoyrup, Jens. 1992. Archimedism, Not Platonism: On a Malleable Ideology of Renaissance Mathematicians. In *Archimede: Mito Tradizione Scienza*, ed. Corrado Dollo, 81–110. Firenze: Leo S. Olschki.

Laird, Walter Roy. 1991. Archimedes Among the Humanists. *Isis* 82(314): 629–638.

———. 2013. Guidobaldo Del Monte and Renaissance Mechanics. In *Guidobaldo Del Monte (1545–1607). Theory and Practice of the Mathematical Disciplines from Urbino to Europe*, ed. Antonio Becchi, Domenico Bertoloni Meli, and Enrico Gamba, 35–51. Berlin: Edition Open Access. http://edition-open-access.de/proceedings/4/3/.

Mach, Ernst. 1919. *The Science of Mechanics: A Critical and Historical Account of Its Development. Translated by Thomas J. McCormack*. 4th ed. Chicago/London: Open Court.

Machamer, Peter. 1998. Galileo's Machines, His Mathematics, and His Experiments. In *The Cambridge Companion to Galileo*, ed. Peter Machamer, 53–79. Cambridge: Cambridge University Press.

Renn, Jürgen, and Peter Damerow. 2012. *The Equilibrium Controversy Guidobaldo Del Monte's Critical Notes on the Mechanics of Jordanus and Benedetti and Their Historical and Conceptual Backgrounds*. Berlin: Edition Open Access.

Stevin, Simon. 1585. *Dialectike ofte Bewysconst*. Leyden: Christoffel Plantijn.

———. 1605. *Mathematicorum Hypomnematum … : Tomus Quartus: De Statica*. Leiden: Ioannis Patius.

———. 1634. *Les Oeuvres Mathematiques de Simon Stevin de Bruges*, ed. Albert Girard. Leiden: Bonaventure & Abraham Elsevier.

———. 1955. *The Principal Works of Simon Stevin. Vol 1: General Introduction. Mechanics*, ed. E. J. Dijksterhuis. Trans. C. Dikshoorn. Amsterdam: Swets & Zeitlinger.

———. 1961. *The Principal Works of Simon Stevin. Vol 3: Astronomy. Navigation*, ed. A. Pannekoek and Ernst Crone. Trans. C. Dikshoorn. Amsterdam: Swets & Zeitlinger.

———. 1964. *The Principal Works of Simon Stevin. Vol 4: The Art of War*, ed. W.H. Schukking. Amsterdam: Swets & Zeitlinger.

———. 1966. *The Principal Works of Simon Stevin. Vol. 5: Engineering. Music. Civic Life*, ed. R.J. Forbes, A.D. Fokker, and A. Romein-Verschoor. Amsterdam: Swets & Zeitlinger.

van Bunge, Wiep. 2001. *From Stevin to Spinoza: An Essay on Philosophy in the Seventeenth-Century Dutch Republic*. Leiden: Brill.

Van Dyck, Maarten. 2006. Gravitating towards Stability: Guidobaldo's Aristotelian-Archimedean Synthesis. *History of Science* 44(4): 373–407.

———. 2013. Argumentandi Modus Huius Scientiae Maxime Proprius.' Guidobaldo's Mechanics and the Question of Mathematical Principles. In *Guidobaldo Del Monte (1545–1607). Theory and Practice of the Mathematical Disciplines from Urbino to Europe*, ed. Antonio Becchi, Domenico Bertoloni Meli, and Enrico Gamba, 9–34. Berlin: Edition Open Access. http://www.edition-open-access.de/proceedings/4/2/.

Chapter 3
Models of Intelligibility in Galileo's Mechanical Science

David Marshall Miller

Abstract Based on an examination of Galileo's mechanics, Peter Machamer and Andrea Woody (and Machamer alone in subsequent articles) proposed the scientific use of what they call *models of intelligibility*. As they define it, a model of intelligibility (MOI) is a concrete phenomenon that guides scientific understanding of problematic cases. This paper extends Machamer and Woody's analysis by elaborating the semantic function of MOIs. MOIs are physical embodiments of theoretical representations. Therefore, they eliminate the interpretive distance between theory and phenomena, creating classes of concrete referents for theoretical concepts. Meanwhile, MOIs also provide evidence for historical analyses of concepts, like 'body' or 'motion', that are otherwise thought to be too basic for explicit explication. These points are illustrated by two examples also drawn from Galileo. First, I show how the introduction of the balance as an MOI leads Galileo to reject the Aristotelian conception of elemental natures. Second, Galileo's rejection of medieval MOIs of circular motion constrains the reference of 'conserved motion' to curvilinear translations, thereby excluding the rotations that had been included in its scope. Both uses of MOIs marked important steps toward modern classical mechanics.

3.1 Introduction

Peter Machamer and Andrea Woody, and Machamer alone in subsequent articles,[1] proposed the theoretical and pedagogical use in science of what they call *models of intelligibility*. As they define it, a model of intelligibility (MOI) is a concrete phenomenon that guides scientific understanding of problematic cases. In particular, Machamer and Woody examine the work of Galileo, and show how he used a

[1] (Machamer and Woody 1994; Machamer 1997; Machamer 1998).

D.M. Miller (✉)
Department of Philosophy and Religious Studies, Iowa State University,
Ames, IA 50011, USA
e-mail: dmm1@iastate.edu

M.P. Adams et al. (eds.), *Eppur si muove: Doing History and Philosophy of Science with Peter Machamer*, The Western Ontario Series in Philosophy of Science 81, DOI 10.1007/978-3-319-52768-0_3

balance as an MOI. Deployed throughout Galileo's corpus, the balance was the primary means by which diverse phenomena were rendered explicable. He explains the behavior of a wide range of physical phenomena by reducing them to equilibrium problems that could in turn be solved by consideration of a suitably configured balance. Machamer and Woody suggest that similar MOIs might be employed throughout the sciences, both in theoretical work and in instructing students.

This chapter, drawing on the same Galilean sources, extends Machamer and Woody's proposal by elaborating the semantic function of MOIs. Scientific theories comprise coordinative definitions that map theoretical concepts onto the phenomena, and MOIs are a way that scientists establish such coordinations. MOIs create classes of concrete referents for theoretical concepts. Beyond facilitating the application of theoretical explanations to problematic cases, MOIs help determine what theories *mean*. Thus, MOIs are not merely productive ancillary heuristics, they are *constitutive* of scientific theories insofar as they designate targets of their application.

Like Machamer and Woody—though for additional reasons—I will suggest that it is historically and philosophically fruitful to recognize the role of MOIs throughout the sciences. In particular, the coordinative function of MOIs makes them historiographically useful. By introducing (or rejecting) MOIs, historical actors alter the application of theoretical explanations to phenomena. That is, they change the theory itself. Therefore, MOIs serve as signposts to loci of theory change, and can be used to track the development of science.

I will illustrate this historiographical usefulness below by examining Galileo's manipulation of MOIs in relation to the reference of 'conserved motion'. By rejecting an MOI commonly accepted by his predecessors, he constrains the applicability of the theoretical concept to curvilinear translations, thereby excluding rotations that had been hitherto included in its scope. This alteration of the coordinative definitions established by MOIs marked an important step toward the advent of the classical inertia theory.

3.2 Coordinative Definitions and Models of Intelligibility

Scientific theories must contain a set of coordinative definitions that mediate between phenomena and theoretical explanations. The argument for this claim is fairly straightforward, at least in general. Phenomena are "out there." They are concrete things moving around in physical space and time.[2] Explanations, however, are "in here." They are inferential relationships between explanatory and empirical

[2] My use of the term 'phenomena' here elides the important distinctions between physical entities, sensory givens, measurements, data, phenomena, and so on. A complete account would venture into territory far afield of the concerns addressed here. Likewise, the reader should not take this use of the term to imply any commitment to scientific realism. One can remain agnostic as to the precise ontological status of phenomena.

concepts, such as causal laws and mental representations of states of affairs.[3] While these cognitive links provide the denotation of a concept, they are (usually) insufficient to pick out which phenomena a theoretical concept refers to. In semantic terms, theoretical explanations can establish the *sense* of a concept via its role in the inferential relationships asserted by the theory, but explanations cannot establish the *reference* of the concept in the phenomena. It follows that the concepts that appear in explanations must be "lined up" with the phenomena in order for the theory to explain anything at all. That is, the conceptual representation of the phenomena that can be derived from the explanatory principles must be applied to the phenomena themselves. The explanation has to get from "in here" to "out there." This is the function of coordinative definitions. They establish the reference of theoretical concepts and determine how the theory represents the world.

To take a simplistic example, suppose that one sets out to explain the fall of a heavy body. Suppose also that one subscribes to modern physics, which explains such phenomena on the basis of gravitational attraction. So one might say, "the body's fall is caused by gravity." That is, the cause 'gravity' brings about 'fall' in a 'body'. Note that the expression 'the body's fall' is not itself a falling body. Rather, it expresses a representation of a phenomenon that connects the concept 'fall' to the concept 'body', and this representation remains to be interpreted. One still has to say, for instance, which phenomena actually *count* as the motion caused by gravity—which phenomena are referred to by the concept 'fall'. For instance, one must specify that the motion of an apple from a tree is a 'fall', while the motion of a ball rolling along a level surface is not. In effect, one has to point to the falling apple and say, "*That* is what I mean by 'fall'." Without such coordination, the theory cannot explain, since its concepts are not associated with any phenomenal entities. Hence, theoretical explanations must be accompanied by coordinative definitions that link theoretical concepts with concrete phenomena. It is an element of the theory that the concept 'fall' is coordinated with certain motions and not others.

This is a widely held view in the philosophy of science, at least in outline.[4] However, discussions of coordinative definitions in the philosophical literature usually occur in contexts of explication or justification. They are offered as components of a reconstruction of a theory in order to show what a theory *actually* says, or even what a theory *should* have said. Philosophers typically do not examine the history of science to find how scientists themselves have gone about establishing coordinations. Yet if they are a necessary part of a theory, as philosophers claim, one

[3] This is not the place to delve into a characterization of scientific explanation. Deductions from law-like principles and boundary conditions can be taken as archetypical.

[4] Coordinative definitions have been viewed as necessary to interpret scientific language under the logical positivists' "received view" of scientific theories and its post-positivist "syntactic view" descendants (sometimes under different names, such as 'correspondence rules'). Many versions of the "semantic view" also presume coordinative definitions to link elements of a theoretical model with the world. The minority exception is that some semantic views rely on isomorphism between a theoretical model and the world to establish the reference of theoretical concepts, without coordinative definitions for each conceptual node in the model. See (Friedman 1999; Friedman 2001; Van Fraassen 2008; Brigandt 2010). See also the discussion in (Miller 2014, 4–6).

should expect to find them in actual scientific practice. One should find scientists saying to each other something like, "*That* is what I mean by 'fall'." Of course, this is not something that happens very often. One does not find scientific papers or textbooks larded with explicit ostentions. Hence the question: how do *scientists* coordinate concepts with phenomena? My claim is that at least some coordinative definitions are hiding in plain sight. They lurk in the examples scientists use to illustrate and communicate their theories. Of course, examples are everywhere in the sciences. However, not just any kind of example works as a coordinative definition. What does fit the bill are Machamer's and Woody's MOIs.

Machamer and Woody introduce MOIs by way of describing Galileo's use of the balance as an explanatory touchstone. Throughout his career, Galileo addressed a variety of explanatory problems regarding bodies floating, falling, and on inclined planes. To provide explanations of these cases, Galileo identifies each target phenomenon with an appropriately weighted balance. Since the behavior of the balance is taken to be well understood, its explanation extends to the problem case. Galileo thus uses the balance as a guide to providing explanations for a range of phenomena. Noting this, Machamer and Woody "propose to call thought directing and problem solving devices such as the balance model *models of intelligibility*."[5] That is, MOIs are phenomena—"devices"—that a scientist takes to be well understood and views as extensible to other, less well-understood cases. Thus defined, MOIs are likely familiar to most scientists and students of science, since they are a kind of exemplary phenomena commonly found in laboratories and classrooms, and described in articles and textbooks. For instance, dissection specimens provide an MOI for, say, 'lungs'. Billiard balls are an MOI for 'collision'.[6]

Machamer and Woody note several properties they take to be characteristic of MOIs. The most important for my purposes is the fact that MOIs have a dual nature: "They are not just abstractions, but can be instantiated by real concrete objects."[7] Thus, on the one hand, an MOI is an abstract, cognitive object. It is a mental representation that utilizes the conceptual apparatus of a theory and can therefore be inserted into the inferential relationships that constitute theoretical explanations. An MOI exists at the conceptual level "in here." On the other hand, an MOI is a concrete object in the physical world. The MOI is itself a phenomenon "out there," which can be interacted with, experienced, remembered, and imagined. Thus, when Galileo refers to the balance in *De Motu*, he is simultaneously referring to an abstract, conceptual representation of a balance and to a concrete object his readers are familiar with—an object they can manipulate either in fact or as recalled in their imagination.

[5] (Machamer and Woody 1994, 219).

[6] In his subsequent papers, Machamer enumerates several additional Galilean examples, including bitumen thrown on a hot iron pan (as an MOI for sunspots), pendulums, water on pavement, and a moving ship (Machamer 1997, 149). For a close analysis of the application of the balance model to materials, see (Biener 2004). For an extensive list of additional examples, see (Meli 2006, 2).

[7] (Machamer and Woody 1994, 221).

Note that this concreteness is what distinguishes MOIs from Tarskian, logical models and other kinds of examples employed in science. For instance, one is often asked to consider a diagram or a thought experiment as part of the explication of some theory. These may indeed serve to direct thought and solve problems, but they are not *per se* part of the phenomena explained by the theory. Rather, they are themselves representations—perhaps physical, as in the case of a diagram, or conceptual, as in a thought experiment—and their reference to phenomena must be determined by additional coordinations that are not already embedded in the example.[8] In the case of MOIs, on the other hand, coordinations *are* embedded, insofar as the conceptual representation of the concrete object is transparent. When one uses an MOI, one stipulates that the physical object is properly conceptualized. The MOI "is accepted as understood or intelligible,"[9] and the two sides of an MOI are simply identified without further interpretation. An MOI is where the conceptual rubber meets the phenomenal road.[10]

For their part, Machamer and Woody emphasize the pedagogical and heuristic functions of MOIs. They show how MOIs can serve as guides for thinking through novel and difficult physical problems. When presented with an unfamiliar case and asked to offer an explanation, one can proceed by analogizing it to an MOI. To do so, one attempts to identify which components of the target case correspond to components of the MOI. When one is successful, the explanation of the behavior of the MOI carries through to the problem phenomenon. As Machamer and Woody put it,

[8] As Machamer and Woody point out, the concreteness of MOIs also distinguishes them from similar suggestions, especially Kuhnian exemplars. For Kuhn, exemplars are problem-solving schema, based perhaps on concrete examples. They can therefore involve MOIs, but they are not themselves MOIs. Exemplars are merely heuristics. They are not explananda. (Machamer and Woody 1994, 219).

[9] (Machamer and Woody 1994, 223).

[10] At this point, one might worry about ambiguities arising from the particularities of an MOI. Imagining a theoretician saying, "by 'conserved motion' I mean *that*" while ostending a ball rolling across a table, one could wonder how the listener would know to ignore the ball's color, etc. But this is not the dialectic I have in mind. MOIs are used against the background of additional theoretical knowledge. Suppose a physics student learning about electrons. After learning what charge and mass and so on are, she learns that an electron has a certain charge and a certain mass. All of this is conceptual. But then the professor stands over a cloud chamber, points to one of spiral tracks and says, "That's an electron." And the student now knows that tracks with that shape are electrons. The representation constituted by the MOI (the cloud chamber track) is *transparent* in the sense that it conflates the concept 'electron' (particle with such and such charge, mass, and so on) with a thing in the world. The track *is* an 'electron' in an unqualified, direct way. Put more bluntly, the electron is an 'electron'. That does not mean, however, that no learning or assumptions were required to get the student to the point of making that conflation. Likewise, when the theoretician says, "by 'conserved motion' I mean *that*," one can assume she is stipulating a certain kind of *motion*, in response to a question like, "which motions count as 'conserved motion'?" The concept of 'motion' itself is antecedently understood (perhaps in light of other coordinations). Moreover, in stipulating the coordination between the abstract mental concept and the concrete object, one is implicitly asserting that the object behaves according to its idealization—in this case, that the ball's motion really is conserved—thereby setting aside accidental properties and imperfections (which might be addressed *post hoc*).

"Every other intelligible instance derives in clear steps, by trial and error or by reasoning, from the intuitively clear case."[11] MOIs are therefore useful heuristics. They simplify the task of applying the theory, and can accelerate discovery and learning.

I have no objection to this account, as far as it goes. But my emphasis on coordinative definitions casts the role of MOIs in a slightly different light and gives them a more fundamental status in the development of scientific theories and explanations. The transparency of an MOI is not so much taken for granted for the sake of solving a problem as *given* as granted to coordinate concepts with the world. MOIs are not merely *exemplary* of the concepts employed in a theory, they are *constitutive* of the theory itself, since they give theoretical concepts phenomenal significance. Moreover, the heuristic and pedagogical functions of MOIs that Machamer and Woody emphasize *derive* from this more fundamental semantic role.[12] MOIs help theorizing and pedagogy because they line up theoretical concepts with the world. When one extends an MOI to a problem case, one is extending the reference of theoretical concepts to corresponding elements of that case—expanding the *meaning* of the theory.

3.3 The Balance as Model of Intelligibility

An example will help bring all of this together. As noted above, Machamer and Woody offer Galileo's appeals to the balance as archetypical uses of an MOI. For the time being, I will focus on only one text, his early *De Motu* essay. This was probably written around 1591, when Galileo was just starting out as a mathematics professor at Pisa. Though never published, it survives as a sequence of manuscript drafts.[13]

De Motu considers "natural motions"—the spontaneous motions of (terrestrial) bodies, like the fall of stones, buoyant bodies rising to the surface of a liquid medium, and balls rolling down inclined planes. At least in the earliest version of the essay, Galileo's theory of natural motion depends on two causal principles:

---

[11] (Machamer and Woody 1994, 223).

[12] Machamer hints at the semantic function of MOIs. In his 1998 paper, he notes that they can serve to "coordinate [a geometrical proof] with experience" (Machamer 1998, 70). Elsewhere, he notes that "mechanical models were a necessary part of the 'proof' or criteria of adequacy for determining a valid explanation" (Machamer 1997, 149). Suggestive as these comments are, they are made in passing, without elaboration.

[13] The manuscripts are preserved in the Biblioteca Nazionale Centrale in Florence as MS 71, available online at http://echo.mpiwg-berlin.mpg.de/content/scientific_revolution/galileo/photographicdocumentation. Favaro published them in Volume 1 of the *Edizione Nazionale* (Galilei 1890–1909, 1:243–419). For translations, see (Galilei 1960; Galilei and Fredette 2000). See (Wallace 1990; Giusti 1998; Fredette 2001) for a discussion of the history and dating of the manuscripts. I am here restricting my attention to the essay forms of *De Motu*, thus ignoring the earliest, dialogue form associated with the set.

weight or heaviness (*gravitas*) and lightness (*levitas*). Indeed, this is how Galileo introduces his project at the opening of the treatise:

> We are going to explain … that all natural motion, whether upward or downward, is the result of the essential heaviness or lightness of the moving body.[14]

In fact, Galileo has inherited these causal principles from Aristotle. As is well known, in Aristotelian physics, earthly and watery bodies possess a natural heaviness, which is an intrinsic tendency downward, toward the center of the universe. Airy and fiery bodies possess lightness, which causes them to rise upward, toward the periphery of the terrestrial realm. Galileo recapitulates this Aristotelian framework in the earliest *De Motu* draft:

> … nature has so arranged it that heavier bodies remain at rest under lighter. We must, therefore now note that bodies which move downward move because of their heaviness, while those that move upward move because of their lightness.[15]

As in Aristotle, terrestrial bodies have two distinct tendencies to move "imposed by nature." To this point in his composition of *De Motu*, then, Galileo retains the Aristotelian conceptual apparatus for explaining natural motions. There are two distinct causal concepts that are independent of one another. Nature places heaviness into some bodies, which causes them to move downward. Likewise, some other bodies are naturally endowed with lightness, and move upward.[16]

But, of course, the concepts discussed in the passage are not a theory of natural motion all by themselves. Without knowing *which* bodies have heaviness and *which* lightness, the concepts explain nothing at all. That is to say, the concepts still require coordinative definitions. Heaviness and lightness must be made to refer to specific features of the phenomena. In Aristotelian physics, heaviness and lightness would each be coordinated with phenomena by their own, separate coordinative definitions, which might take the form of an MOI. Thus, heaviness was often linked to stones, and their fall when released was the referent for the effect of heaviness.[17] An Aristotelian might drop a stone and say, in effect, "That is heaviness." On the contrary, lightness was coordinated with smoke or fire, and the rise with its effect. "That is lightness," the Aristotelian might say about the flame and smoke rising from a burning candle.[18]

However, in the fifth chapter of *De Motu*, Galileo introduces a new, mechanical MOI, the balance, as a way to coordinate the concepts with the phenomena:

> I propose … to reduce the matter to a consideration of the balance and to explain the analogy (*convenientia*) that holds between bodies that move naturally and the weights of the

[14](Galilei 1960, 13; Galilei 1890–1909, 1:251).

[15](Galilei 1960, 16).

[16]Machamer has written on this subject, too. See (Machamer 1978).

[17]Aristotle himself refers to the motions of wood, bronze, gold, and lead; e.g., *De Caelo*, IV.1–2, 308a6–308b10.

[18]Aristotle offers fire, vapor, and smoke as examples of light bodies; e.g., *Meteorology* I.4, (341b1–341b17). In *De Motu*, Galileo devotes a chapter to refuting Aristotle's ascription of lightness to fire (Galilei 1960, 55–61).

> balance. My aim is a richer comprehension of the matters under discussion, and a more precise understanding on the part of my readers.[19]

Galileo sets out to explain natural motions by connecting them to the behavior of a balance, which is a concrete object with which Galileo assumes his reader is familiar. This, Galileo supposes, will assist the reader's comprehension of his explanations. Yet the balance can only be an aid to comprehension if it does not itself stand in need of interpretation. Galileo is assuming the interpretive transparency of the MOI. The balance will unproblematically instantiate his explanatory concepts.

Consider Galileo's ensuing discussion of the equal-arm balance:

> Let line *ab* (in Fig. 3.1), then, represent a balance, whose center, over which motion may take place, is the point *c* bisecting line *ab*. And let two weights, *e* and *o*, be suspended from the points *a* and *b*. Now in the case of weight *e* there are three possibilities: it may either be at rest, or move upward, or move downward. Thus if weight *e* is heavier than weight *o*, then *e* will move downward. But if *e* is less heavy, it will, of course, move upward, and not because it does not have weight, but because the weight of *o* is greater.[20]

The behavior of a balance comes as no surprise; Galileo says it is a matter "of course" (*profecto*). Anyone familiar with the operation of a balance knows—and can imagine—that the heavier-weighted arm will descend, the lighter-weighted will ascend, and the balance will remain in equilibrium if the arms are equally weighted.

But notice also how the balance functions as a coordinative definition. One concept in Galileo's explanatory theory is 'heaviness'. To what does this concept refer? The balance *shows* us: 'heaviness' *is* a weight suspended from a balance. 'Heaviness' in the abstract *just is* heaviness in the concrete. Pointing at a balance, Galileo can say, "*That* is heaviness." Moreover, heaviness is what makes the concrete balance actually move. So *that* motion in the phenomena is explained by the concept 'heaviness' in the theory.

As Galileo himself acknowledges, the introduction of the balance as an MOI yields a significant departure from his predecessors' Aristotelian explanations. Notice that, in Galileo's account of the phenomena, the cause of all three possible states of a body (stasis, moving up, moving down) is *heaviness*—the natural tendency of a terrestrial body to descend:

> From this it is clear that, in the case of the balance, motion upward as well as motion downward takes place because of weight, but in a different way. For motion upward will occur for *e* (in Fig. 3.1) on account of the weight of *o*, but motion downward on account of its own weight.[21]

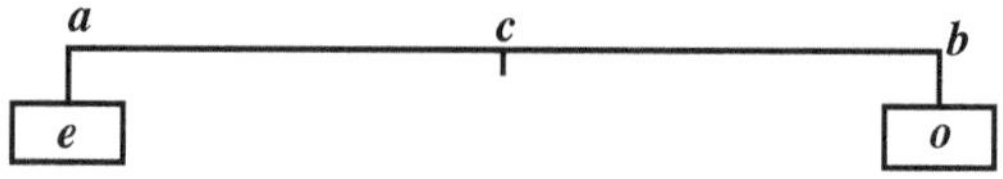

Fig. 3.1 Equal-armed balance in *De Motu* (Galilei 1890–1909, 1:257)

[19] (Galilei 1960, 20).

[20] (Galilei 1960, 21).

[21] (Galilei 1960, 21).

Both bodies on a balance have natural heaviness. Upward motion, where it occurs, results from the greater heaviness of the other body. Lightness does not enter into it, which is to say that the concept has no direct phenomenal reference. There is no feature of the phenomenon to which Galileo might point and say, "That is lightness." But all of this is clear from the familiar behavior of a balance. The reader knows, because she has interacted with a real, concrete balance, that anything without a counterbalance will move *down* because of its heaviness. It is only when the counterbalance has greater heaviness that an object moves up.

Hence, Galileo's use of the balance as an MOI reconfigures the Aristotelian coordinations of 'heaviness' and 'lightness'. It revokes the coordination between spontaneous upward motion and lightness, and attributes upward motions to the action of a heavier medium instead:

> Besides, in the case of bodies moving naturally, as in the weights in a balance, the cause of all motions, up as well as down, can be referred to weight alone. … And from this one can easily understand (contrary to Aristotle *De Caelo* I.89) that what moves [up] moves, as it were, by force and by the extruding action of the medium.[22]

All natural motions, up and down, are the effect of heaviness alone, as one may "easily understand" from a balance. In fact, this is a genuinely novel theory of natural motions. As Galileo himself recognizes, upward motions are explained in a new manner, "contrary to Aristotle."

The fact that the use of the balance as an MOI changes the theory clinches the case. If a new MOI can change the content of a theory, then MOIs must be part of the theory. It is worth emphasizing, however, exactly where the novelty lies. It is not at the level of explanatory concepts. Galileo's theory does not employ a new concept of heaviness or lightness. For Galileo as for Aristotle, heaviness is an intrinsic causal power that brings about the downward motion (or tendency thereto) of terrestrial bodies and lightness is its opposite. Indeed, in the passage quoted above, Galileo provides a perfectly Aristotelian definition of lightness. Where Galileo departs from Aristotle is in his coordination of these concepts with the phenomena. We can imagine an Aristotelian pointing at a rising body and saying "*That* is lightness." Galileo responds, "No. *That* is the heaviness of the medium." This is because (and now I am quoting) "the motion of bodies moving naturally can be suitably reduced to the motion of weights in a balance."[23] Lightness no longer has phenomenal significance for natural motion. It does not refer to any independent feature of nature.

A concept without reference, of course, is not explanatorily useful, and can be excised from a theory. As one might expect, this is Galileo's next move. The second version of the *De Motu* essay eliminates lightness as an independent and distinct causal principle, and redefines the term as mere "deficiency of heaviness." Thus, the new opening begins:

---

[22] (Galilei 1960, 22–23).

[23] (Galilei 1960, 23).

> As we will explain later that all natural motion is the result of an excess or a deficiency of heaviness ...[24]

Compare this to the "essential heaviness or lightness" in the original start of the treatise quoted above. Subsequent revisions of *De Motu* also serve to stress that heaviness is the sole causal principle. The third version, for example, adds a chapter titled "Proof that upward motion cannot, on the part of the moving body, be natural,"[25] which concludes:

> There is a single matter in all bodies, and it is heavy in all of them. But that same weight cannot possess natural inclinations that are contrary. Therefore, if there is one natural inclination, the opposite must be contrary to nature. But the natural inclination of weight is toward the center. Therefore it follows that the inclination away from the center is contrary to nature.[26]

The *only* natural source of terrestrial motion is heaviness, and only spontaneous downward motion is natural. The contrast with Aristotelian theory could not be starker.

So, by the time he set *De Motu* aside, Galileo had completely replaced the Aristotelian theory of natural motion at both conceptual and coordinative levels. He unified explanations by supposing only heaviness, instead of heaviness and lightness. But Galileo's adoption of his new theory was neither sudden nor mysterious. The crucial use of the balance as an MOI first altered the coordinative definitions in the theory, and this led naturally to the reconceptualization of lightness as a dependent and relative notion.[27]

3.4 Models of Intelligibility in Historiography

This discussion of Galileo's balance model in *De Motu* leads to a historiographical point. Introduction—or elimination—of an MOI is a mode of scientific change. By altering the coordination between theory and the world, an MOI changes the meaning of the theory. So by attending to the use of MOIs by historical actors, historians of science can develop accounts of the development of scientific theories. An examination of MOIs illuminates the historical meanings of important concepts and shows how they have changed over time. For instance, Galileo's use of the balance eventually allowed him to introduce the novel and important concept of *momento*, or positional heaviness—heaviness actually exerted by a partially constrained body.[28] More importantly, in my view, this method is particularly useful to elucidate

---

[24] (Galilei and Fredette 2000, 91, slightly altered; Galilei 1890–1909, 1:341).

[25] (Galilei 1960, 118).

[26] (Galilei 1960, 120).

[27] Elsewhere I have described such alternating shifts in the conceptual and coordinative components of a theory 'reciprocal iteration'. See especially (Miller 2011) and (Miller 2014, ch. 1).

[28] See (Machamer 1998, 61).

basic concepts, such as 'body' or 'motion', complete explications of which one would not expect an author to provide.[29] Yet, what counts as a 'body' or 'motion' have changed in fundamental ways that have deeply altered the course of science. MOIs help to track those changes.

Consider another example drawn from Galileo's mechanics: Galileo's concept of 'motion'. One of the many achievements for which Galileo is justly famous is his theory of inertia. This holds that motion (and rest) is a state of a body that is conserved unless some force acts upon it. For example, a heavy body on a smooth horizontal surface is not subject to the action of any force, since its intrinsic heaviness is counteracted by the extrinsic support of the plane. As a result, the body, once moved, will continue to move uniformly:

> And therefore, with all the external impediments removed, a heavy body on the spherical surface concentric to the Earth will be indifferent to rest and to movement toward any part of the horizon, and it will remain in the state in which it has been put; that is, if it has been put in a state of rest it will remain in it, and if it has been put in motion, toward the west, for example, it will remain in the same motion.[30]

Without the action of any force, the body will be "indifferent" to its horizontal motion, and will conserve that state in each moment.

There is a question of coordination here, however. Which phenomena do Galileo's theory of "indifference" or, as we would say, inertia explain? Specifically, what are we to take as "the same motion" conserved by a body upon which no force acts? For one thing, as is well known and can be seen in the above passage, Galileo supposes that *circular* motions are conserved. The horizontal motion of the body would encircle the earth. So "the same motion" follows a circular trajectory. Yet there is a further question to be asked. A body's circular motion can be counted in several ways. One might say that the body subtends equal central angles in equal times, or one might say that the body subtends equal curvilinear distances in equal times. The "motion" that is the same from moment to moment might be an angular rotation or a curvilinear translation. So which is it? Which kind of circular motion is coordinated with the conservation principle in Galileo's inertial theory? What does his theory *mean*?

This question has historical importance. In classical, Newtonian physics, inertial 'motion' refers specifically to linear translations. The 'motion' conserved from moment to moment is a translation along a straight line. Angular rotations, on the other hand, are *not* fundamentally inertial. Strictly speaking, each part of a rotating object does not undergo one single motion, but an infinite number—a motion in every direction in the plane. Continued rotation is only due to the constraint on each part from the cohesion of the object.[31] This view is continuous with one version of Galilean inertia, but not with the other. Angular rotations are not commensurate

---

[29] Though they do not put it in such terms, (Machamer and Hepburn 2004) show how Galileo uses a pendulum as an MOI for the concept 'time'.

[30] (Galilei 1890–1909, 5:134; Galilei and Scheiner 2010, 125, translation slightly altered).

[31] See the elaboration of Newton's First Law of Motion: "A spinning hoop [or top; *trochus*], which has parts that by their cohesion continually draw one another back from rectilinear motions, does

with rectilinear translations, but translations along curvilinear trajectories need only be "straightened out" in order to constitute the modern referent of inertial motion. Hence, if one is curious about the advent of the modern inertial theory, one should be curious about how curvilinear translations in particular came to be the referents of conserved motion. In fact, one can satisfy that curiosity by examining the MOIs theorists have used.

Galileo's predecessors did not distinguish angular rotations and linear translations in their theories of motion. Writing in the fourteenth century, for instance, John Buridan argued against the Aristotelian view that the medium is what causes the continued motion of a body. To do so, he offers some "experiences" as MOIs for motion:

> The first experience concerns the top (*trocus*) and the smith's mill (i.e., wheel—*mola fabri*) which are moved for a long time and yet do not leave their places. … The second experience is this: A lance having a conical posterior as sharp as its anterior would be moved after projection just as swiftly as it would be without a sharp conical posterior.[32]

The fact that the top or smith's wheel does not change place and that the lance has a sharp posterior end show that the medium, i.e., the air, cannot be the cause of the continued motion. In Buridan's own view, the moving agent imparts an *impetus* upon the moving body, which then persists in the body until removed by some counteracting agent. What is important here, though, is the fact that the motion in the case of the top and wheel are angular rotations, while the motion of the lance is a linear translation, be it straight or curved. The motion that continues "for a long time" is the rotation of the whole top or wheel, while the lance does not rotate at all. Buridan's use of these MOIs shows that, for him, 'motion' can refer to either an angular rotation or a linear translation. The concept in the explanatory theory is coordinated with both kinds of phenomena. Moreover, Buridan's "experiences" were standard examples of 'motion' throughout the medieval period. They appeared in several authors down to Galileo and beyond, including John of Holland, Nicole Oresme, Albert of Saxony, Erasmus Rheinhold, and Domingo De Soto.[33] Meanwhile, the consistency of examples across several centuries indicates a longstanding consensus about what the concept of 'motion' was *supposed* to refer to.

Returning to Galileo, one finds in *De Motu* that Galileo at first follows his predecessors in coordinating 'motion' with *both* angular rotations and curvilinear translations. This is again made clear by the MOIs he offers. In *De Motu*, Galileo attempts to show that there are circumstances in which all the forces acting on a body are counterbalanced by opposing forces. Under these conditions, the body will be "indifferent" to motion, such that it will not resist motion and might be moved "by the very smallest force, indeed, by a force less than any given force."[34]

---

not cease to rotate, except insofar as it is retarded by the air" (Newton 1999, 416). Note that Newton uses a standard example, the *trochus*, discussed below.

[32] (Clagett 1959, 533).

[33] (Wallace 1968; Westman 1980, 113).

[34] (Galilei 1960, 65–66).

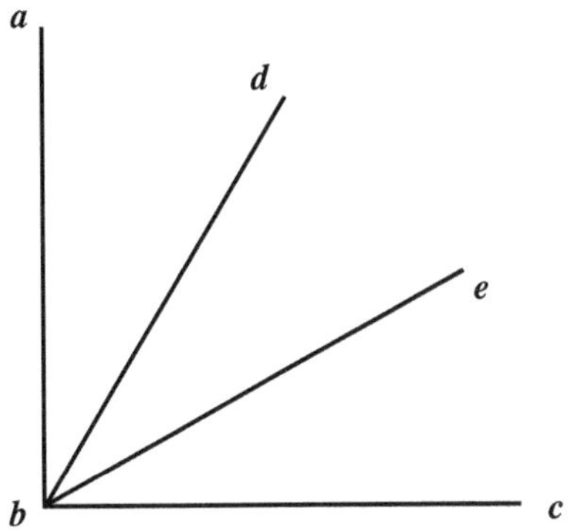

Fig. 3.2 Inclined planes in *De Motu* (Galilei 1890–1909, 1:296)

To exemplify such motion, Galileo first discusses bodies on inclined and horizontal planes. Here, he supposes that the bodies traverse linear lengths of planes:

> Now from point *b* (in Fig. 3.2) let any number of lines be drawn making acute angles with line *bc*, e.g., lines *bd* and *be*. The problem, then, is why a body moving down descends most quickly on line *ab*; and on line *bd* more quickly than on *be*, but more slowly than on *ba*; and on *be* more slowly than on *bd*.[35]

Here, the 'motion' the body has along each plane is a linear translation. Likewise, the motion a body would have along the horizontal *bc* would be "indifferent" to that linear translation. Though, again, the horizontal plane is really a spherical surface. So indifferent motion is here a curvilinear translation.

Yet in the very same text, just a few pages later, Galileo also employs a rotating sphere as an MOI for motion:

> Now if the body is of homogeneous parts, e.g., a marble sphere, rotating on an axis, such motion will be neither natural nor forced. … the sphere is always in equilibrium.[36]

Since the sphere is in equilibrium, no forces will act on it, and it will also be indifferent to motion, just like the body on the horizontal plane. But the "motion" in this case is the rotation of the whole sphere, not the curvilinear motions of its various parts. This MOI coordinates indifferent motion with an angular rotation. So, early in his career, Galileo follows his predecessors in adopting an application of the concept of motion that does not discriminate between translations and rotations. The inclined plane coordinates motion with a linear translation. The marble sphere coordinates it with an angular rotation.

At the end of his career, however, Galileo withdraws one of these MOIs. In the *Discourses and Mathematical Demonstrations Concerning Two New Sciences*, published in 1638, Galileo revises the mechanics he first developed in *De Motu* a half-century before. Once again, he discusses the motion of bodies on planes. As before, he argues that a body on a horizontal surface is indifferent to motion:

[35] (Galilei 1960, 63–64).

[36] (Galilei 1960, 75).

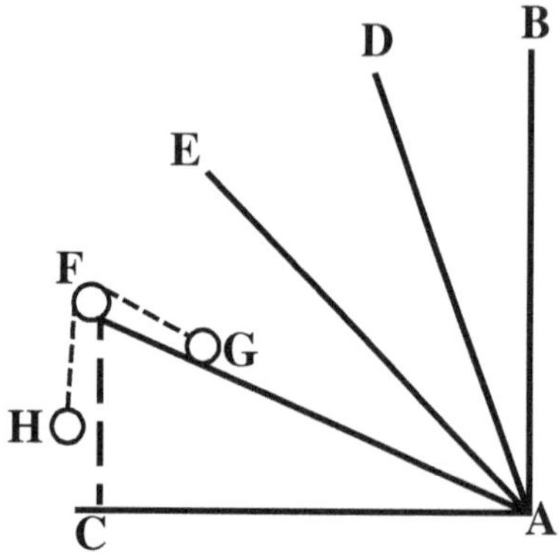

Fig. 3.3 Inclined planes in *Discourses* (Galilei 1890–1909, 8:215)

> I say that the impetus of the heavy body for descending is maximal and total along the vertical *BA* (in Fig. 3.3), … and is finally completely extinguished on the horizontal *CA*, where the moveable is found to be indifferent to motion and to rest, and has in itself no inclination to move in any direction, nor yet any resistance to being moved.[37]

Galileo concludes that a body on the horizontal surface is not subject to any action, and is indifferent to motion and rest. But as we have seen, motions along planes are linear translations, not angular rotations. So translations are the focal referents for 'motion'. Rotations, on the other hand, do not appear in the *Discourses* in contexts where Galileo discusses the physics of local motion.[38] A rotating body, like a marble sphere or child's top, is no longer an MOI. By 'motion', Galileo *just* means linear translations.

This restriction of the reference of 'motion' to linear translation culminates in Galileo's famous derivation of the parabolic trajectory of projectiles.

> Imagine a horizontal line or plane *ab* (in Fig. 3.4) situated on high, upon which the moveable is carried from *a* to *b* in equable motion, but at *b* lacks support from the plane….[39]

Here, "equable motion" refers specifically to the linear translation of the ball on the horizontal plane. Since the plane is horizontal, the motion is conserved. Once it runs off the edge, a vertical acceleration is added. The resulting composed motion, Galileo shows, is parabolic—or at least approximately so. The horizontal motion is not really straight, but along a circle concentric with the earth. Galileo says that for the small-scale phenomenon in question, the difference is negligible. However, this approximation only works if Galileo assumes that the motion is a linear translation. This is the only way in which the straight motion can approximate a curvilinear motion. Straight motion approximates curvilinear motion, but not rotation.

---

[37] (Galilei 1989, 172).

[38] Galileo does discuss the rotation of the *rota aristotelica* in Day 1, but the subject there is vacua in material substance, not mechanics. (Galilei 1989, 29–34).

[39] (Galilei 1989, 221).

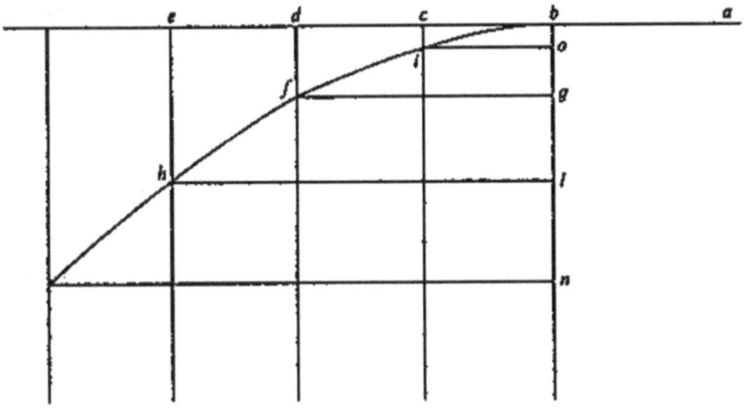

Fig. 3.4 Parabolic trajectory in *Discourses* (Galilei 1890–1909, 8:272)

Eliminating the marble sphere as an MOI leads to the specific coordination of 'motion' with linear translations. This changes the theory, bringing it closer to the modern view, thereby paving the way for the modern theory of inertia. Tracking the use of MOIs helps us identify this key move in the history of mechanics, even in the absence of explicit definitions of 'motion'. Of course, there is much more to the story,[40] but what has been said shows how MOIs can guide historical investigations of science.

3.5 Conclusion

Theories necessarily contain coordinations between their concepts and the phenomena. In actual scientific practice, one way those coordinations are established is by the use of *models of intelligibility*. These are a theory's windows onto the world. And since they are a necessary part of a theory, MOIs are historically useful. Just as they guide scientists' understanding of the phenomena, they can guide historians' understanding of the science.

Acknowledgments My deepest gratitude to Peter Machamer, for letting me ask the questions I wanted to ask, and teaching me how to seek the answers. Thanks also to the editors of this volume, especially Adams and Biener, whose comments and suggestions improved this paper immensely.

[40] See (Miller 2014, ch. 5).

References

Biener, Zvi. 2004. Galileo's First New Science: The Science of Matter. *Perspectives on Science* 12(3): 262–287.

Brigandt, Ingo. 2010. The Epistemic Goal of a Concept: Accounting for the Rationality of Semantic Change and Variation. *Synthese* 177(1): 19–40.

Clagett, Marshall. 1959. *The Science of Mechanics in the Middle Ages*. Madison: University of Wisconsin Press.

Fredette, Raymond. 2001. Galileo's *De Motu Antiquiora*: Notes for a Reappraisal. In José Montesinos and Carlos Solís (eds.), *Largo Campo di Filosofare: Eurosymposium Galileo 2001*, La Orotava: Fundación Canaria Orotava de Historia de la Ciencia, 165–181.

Friedman, Michael. 1999. Geometry, Convention, and the Relativized A Priori: Reichenbach, Schlick, and Carnap. In *Reconsidering Logical Positivism*, Cambridge: Cambridge University Press, 59–70.

Friedman, Michael. 2001. *Dynamics of Reason*. Stanford: CSLI Publications.

Galilei, Galileo. 1890–1909. *Le Opere di Galileo Galilei*. Edited by Antonio Favaro. 20 vols. Florence: Barbera.

Galilei, Galileo. 1960. *On Motion and On Mechanics*. Translated by I. E. Drabkin and Stillman Drake. Madison: University of Wisconsin Press.

Galilei, Galileo. 1989. *Two New Sciences*. Translated by Stillman Drake. Toronto: Wall & Emerson.

Galilei, Galileo, and Raymond Fredette. 2000. De Motu Antiquiora. Available from http://echo.mpiwgberlin.mpg.de/content/scientific_revolution/galileo/englishtranslation.

Galilei, Galileo, and Christoph Scheiner. 2010. *On Sunspots*. Translated by Eileen Reeves and Albert Van Helden. Chicago: University of Chicago Press.

Giusti, Enrico. 1998. Elements for the Relative Chronology of Galilei's *De Motu Antiquiora*. *Nuncius* 13(2): 427–461.

Machamer, Peter. 1978. Aristotle on Natural Place and Natural Motion. *Isis* 69(3): 377–387.

———. 1997. Comment: A New Way of Seeing Galileo's Sunspots (and New Ways to Talk Too). In Henry Krips, J. E. McGuire and Trevor Melia (eds.), *Science, Reason, and Rhetoric*, Pittsburgh: University of Pittsburgh Press, 145–152.

———. 1998. Galileo's Machines, His Mathematics, and His Experiments. In Peter Machamer (ed.), *The Cambridge Companion to Galileo*, Cambridge: Cambridge University Press, 53–79.

Machamer, Peter, and Andrea Woody. 1994. A Model of Intelligibility in Science: Using Galileo's Balance as a Model for Understanding the Motion of Bodies. *Science & Education* 3(3): 215–244.

Machamer, Peter, and Brian Hepburn. 2004. Galileo and the Pendulum: Latching on to Time. *Science & Education* 13: 333–347.

Meli, Domenico Bertoloni. 2006. *Thinking with Objects: The Transformation of Mechanics in the Seventeenth Century*. Baltimore: Johns Hopkins University Press.

Miller, David Marshall. 2011. Friedman, Galileo, and Reciprocal Iteration. *Philosophy of Science* 78(5): 1293–1305.

———. 2014. *Representing Space in the Scientific Revolution*. Cambridge: Cambridge University Press.

Newton, Isaac. 1999. *The Principia: Mathematical Principles of Natural Philosophy*. Translated by I. Bernard Cohen and Anne Whitman. Berkeley: University of California Press.

Van Fraassen, Bas C. 2008. *Scientific Representation: Paradoxes of Perspective*. Oxford: Oxford University Press.

Wallace, William A. 1968. The Enigma of Domingo de Soto: *Uniformiter difformis* and Falling Bodies in Late Medieval Physics. *Isis* 59(4): 384–401.

———. 1990. The Dating and Significance of Galileo's Pisan Manuscripts. In Trevor Harvey Levere and William R. Shea (eds.), *Nature, Experiment, and the Sciences*, Dordrecht: Kluwer Academic Publishers, 3–50.

Westman, Robert S. 1980. The Astronomer's Role in the Sixteenth Century: A Preliminary Study. *History of Science* 18(2): 105–147.

Chapter 4
Equilibrium, Natural Motion, and Models of Explanation

Brian Hepburn

Abstract A key theme in the historiographical work of Machamer has been the ways that motion is made intelligible through explanatory means of natural motion and models of the simple machines such as the lever and pendulum. One way of spelling out the explanatory value of these strategies is through the concept of equilibrium. Natural motion and simple machines allow the simplification of complex problems in terms of self-evident, intelligible equilibrium conditions. This chapter connects the theme of equilibrium and natural motion across Machamer's work on mechanisms and mechanical explanation, on Aristotle, Galileo, Descartes and Newton, and on the pendulum and mental models. Just as equilibrium can be found within science, it also becomes a model of intelligibility for doing history and philosophy of science: a normative but objective representation of the important properties of science.

4.1 Introduction

In the spring of 2001, Prof. Machamer set the final exam for the history core course for the students coming in to the History and Philosophy of Science program at the University of Pittsburgh. In studying for that exam, I had the intuition that the concept of natural motion would be important. Although it was never explicitly identified as a theme in the course, Machamer had managed to plant the idea that the concept could be very useful for understanding developments and differences in science. As it turned out, natural motion was a long-essay topic on the exam.

Natural motion, equilibrium, and the simple machines are variations of the same explanatory strategy: bracketing-off aspects of change as normal and so not requiring of explanation. It is as much a strategy of explaining *away* as it is one of explain-

---

All things are not possible at all times [Machamer (1975), 381].

B. Hepburn (✉)
Department of Philosophy, Wichita State University, Wichita, KS 67260, USA
e-mail: brian.hepburn@wichita.edu

M.P. Adams et al. (eds.), *Eppur si muove: Doing History and Philosophy of Science with Peter Machamer*, The Western Ontario Series in Philosophy of Science 81, DOI 10.1007/978-3-319-52768-0_4

ing. Once some motion is connected with the nature of an entity it no longer makes sense to ask "why?" One can no more ask why some natural property of X belongs to X than one can ask why X is an X? Similarly, what would otherwise be a complicated or even intractable phenomenon can be made manageable by discovering underlying equilibrium conditions. Changes which "cancel out" can be ignored. Parts of systems in equilibrium can be "black-boxed", the internal workings serving to translate an input to an output. Mechanisms can be identified in this way.

The ideas of natural motion, equilibrium, and mechanism are thus interconnected, but without paying close attention to the constraints offered by the interconnections, the overall idea can appear trivial. Regarding natural motion in particular, conceptions have gone from the cosmologically ordered natural motions of the elements, to natural circular motion around mechanical centers, to inertial straight-line motion, to paths of least action, and many other varieties along the way. Along with adopting some such conception of natural motion, there is still plenty to be done working out the best account of that natural motion: its more technical description (up, down, accelerated, conserved, dependent on what?), how to mathematize it, how to use it to explain other phenomena. A conception of natural motion can provide the foundation of an explanatory framework, but constructing the fuller account, capable of explaining a phenomena other than natural motion itself, requires a network of constraints and connections among concepts about things and their properties and behaviours. One way for the network to provide explanations is by reading the connections as mechanical interactions, revealing one behaviour or change in property to be in equilibrium with (balanced out by) some other or others. So the question of whether some property should be thought of as a natural property is neither trivial in its execution nor its implications.

Machamer is a well-known proponent of the mechanistic account of explanation, and of the basic intelligibility of the simple machines. Natural motion and equilibrium are less well-known, and less explicit, in Machamer's work. A recent paper co-authored with James McGuire and Hylarie Kochiras does draw connections between mechanical explanation and in arguing for a general unified conception of Mechanical Philosophy— one which, as a case-study in that paper, can accommodate Newtonian attractionism. The doors may be flung too wide, however, resulting in a trivializing of the applicability of the category and so a loss of explanatory insight.

At the outset, then, I will take a closer look at this recent paper and its potential problems. To mitigate those problems, I offer an analysis of the role of natural motion in mechanical philosophy, completing the mechanism-equilibrium-natural motion trinity for the Machamer, McGuire, Kochiras paper. Mechanical philosophy based in natural motion should be understood as primarily about explanation, rather than an account of physical reality. Ironically, I will argue that Machamer's account of explanation would benefit from a resistance to the over ontologizing Machamer and McGuire decry with respect to Descartes in *Descartes's Changing Mind* ([Machamer and J.E. (2009)]). This emphasis on explanation and understanding, as a model of intelligibility versus a model of the world in actuality, provides an avenue for traversing much of Machamer's work. So this paper also aims to provide a survey of some of those intersections. Free from over-ontologizing, these explanations can reveal to the historian and philosopher of science a rich, pluralistic, tapestry of sciences, both historical and contemporary.

4.2 Newton: The Mechanical Philosophy

In their 2012 paper, Machamer, McGuire, and Kochiras (hereafter, MMK) argue for what amounts to a reboot of Newtonian scholarship [Machameret al. (2012)]. Newton allows for — or is at least agnostic about — the possibility of action at a distance (gravitational attraction, in particular). Therefore, he would seem not to belong to the tradition of mechanical philosophy as standardly understood. While this tradition can be identified with a plurality of overlapping but distinct views (e.g. [Gabbey (2002)]), they all seem to share a commitment to efficient causation through contact only. This is what the standard historiography holds (e.g. [Janiak (2008)].) MMK want to challenge this essential characteristic and, in so doing, bring Newton within the fold of a unified tradition of Mechanical Philosophy (their capitalization) more broadly understood.

Their attempt to broaden the notion of "mechanical philosophy", however, threatens to dissolve the distinction entirely. It opens the door to identifying as a mechanical explanation any explanation of natural phenomena which employs something like a balance or equivalence between cause and effect. Thus, Aristotle becomes a mechanical philosopher (e.g., effect is proportional to force); Renaissance magicians become mechanical philosophers (the balance between macro and microcosm); God is a mechanician, no matter what the theology, so long as He/She never acts superfluously. However, I come to fest Machamer, not to bury him (et al.). And I think there is a way out. It is to bring to the fore a concept which underlies much of Machamer's historiography and analysis of the sciences: the concept of equilibrium. Equilibrium is an explanatory concept which connects natural motions together, primarily as mechanisms. All three components (equilibrium, explanation, natural motion) are required for a proper mechanical philosophy.

The first move made by MMK against pluralism and in favour of unified Mechanical Philosophy is to reject the idea that mechanical philosophy and mechanics were distinct concepts, even for the historical actors who used them [Machamer et al. (2012), 371]. Gabbey (2002) has studied how these terms were used in the early modern period and proposed a taxonomy according to which one term, mechanics, refers to those manual activities using tools and machines but without any strong theories attached to them. This was distinct from the second term, mechanical, or mechanical philosophy, which meant the more reflective study of the operation and usage of machines. MMK want to point to an even higher-level characterization, according to which both kinds of usage are "somehow referring to versions of a general Mechanical Philosophy…" [Machamer et al. (2012), 371].

MMK therefore propose a novel characterization of the Mechanical Philosophy and they identify two traditions as feeding in to it. One tradition runs through the Aristotelian mixed sciences, and Galileo, Boyle, Descartes, embracing quantification and mathematical laws, algebra and geometry. The other is the Archimedean simple machines, which MMK see as one tradition encompassing both mechanics and the mechanical, both theory and practice. In the paper they identify the Mechanical Philosophy as the hybrid of these two traditions and argue that Newton belongs to that tradition. At the same time, identifying the traces of the two traditions within Newton's work also provides support for MMK's unification thesis. It's a difficult bit of manoeuvring: the tradition exists and Newton belongs to it, and that Newton belongs to it is evidence that the tradition exists. This is a benign circularity, though, and a part of a consistent, self-supporting reading of the history of science. The value of the reading lies in the sense it helps us to make of developments in natural philosophy during the Scientific Revolution but also, importantly, during the often neglected eighteenth-century after. The greater worry, to be addressed in a moment, is that the new view is too encompassing.

The argument to establish Newton as a representative of Mechanical Philosophy turns on a couple of key premises. One is the identification of law explanations with expressions of balance or equilibrium. If laws, such as Newton's 2nd Law, or the law of universal gravitation, are understood by Newton to describe a balance (or some machine) then Newton belongs to the expanded Mechanical Philosophy tradition. In short, mathematical laws are new ways of representing those models of intelligibility that are the simple machines. "In what follows, we aim to show two things: first, that Newton placed himself in this tradition of the simple machines and, second, that Newton used the balance as the model for treating gravitational action as a simple machine." [376] There are references to the simple machines (balance, lever, pendulum) throughout *Principia*, and MMK show convincingly how the preface demonstrates Newton's (self-acknowledged) indebtedness to the mechanical tradition. The question is whether the way in which Newton goes beyond that indebtedness represents breaking with the tradition or extending it. Since Newton seems to allow that possibly gravity can act at a distance, without contact, and he rejects the leading mechanical explanation of planetary dynamics, namely, the ether-vortices theory, Newton is a tradition breaker if contact action is a sine qua non of mechanical philosophy. MMK do not think that it is.

> We should, in fact, resist reading Newton along purely Boylean lines, according to which the Mechanical Philosophy is intertwined with the corpuscular philosophy and adherence to contact action, and consider Newton's Mechanical Philosophy against the backdrop of two other traditions that affected the way in which seventeenth-century natural philosophers thought of motion as the principal efficient cause acting in nature: the mixed sciences and the simple machines [Machamer et al. (2012)].

What they do take to be a sine qua non of seventeenth century mechanics is the "study of the mathematics of the simple machines" [Machamer et al. (2012), 383]. A key element of the mathematics of the machines is that it represents —whether through ratios, geometry, algebra, or differential calculus —equilibrium relations among physical quantities. This is more key, more essential, than interpreting the mathematics as representing contact action. Interaction, yes, but why contact necessarily?

> The element Newton draws from the simple machine tradition is his means of discovering or identifying genuine causes in the world....It is therefore legitimate to understand

> Newton's natural philosophy as mechanical, in virtue of the mathematical approach that he draws from the simple machine tradition and takes to identify forces as physical causes [Machamer et al. (2012), 382].

On this point it seems a matter of (historiographical) preference. One might answer the rhetorically posed question "why contact necessarily?" by asserting that contact action is what makes it a mechanical philosophy. Causes and interactions among entities might be understood in many ways, even as efficient causation, but the point of the mechanical label is to identify those which emphasize contact efficient causation.

The preference of MMK is rather to emphasize motion as a cause and expressed in mathematical laws, regardless of whether contact is behind those laws or not. A rejection of Aristotelianism was an ingredient of the mechanical philosophy; so, too, were new unified matter theories [Machamer et al. (2012), 373]. But "even more important" was the idea that causation could be captured through regularities of motions, specifically quantitative laws. However, not just any laws will do, but laws descriptive of, or interpretable on models of, the simple machines — and what characterizes the mathematics of the simple machines, in turn, is that they describe equilibrium conditions among natural motions of the parts of the machines. Levers operate through "balanced reconfigurations of the opposing ends", for example, and thus

> …motion could be conceived in terms of equilibria relations, as somehow resulting in some sort of stasis (its later connection to statics explicable by the fact that we can measure proportions only by treating them as static, even though they may be continuously changing). And rest could be conceived as a stable equilibrium state.

This point, that equilibrium could provide a mechanical explanation of non-static interactions, is important for justifying the MMK story. Because dynamic equilibrium turns out to be such an important part of mechanics in the eighteenth-century— much more so than contact action — the MMK reconstruction of the tradition of Mechanical Philosophy is more continuous with continuing developments in that period. On the standard story, which cleaves to contact action, mechanical philosophizing ends, more or less, with Newton.

The eighteenth-century provided an astounding demonstration of how concepts like the balance, lever, and equilibrium could be powerfully wielded in solving problems in natural philosophy (primarily by re-defining the problems, and natural philosophy.) They did so in the hands of mechanical philosophers like Hermann, the Bernoullis, d'Alembert, Euler, and Lagrange. The part of the Scientific Revolution which was the mathematization of Natural Philosophy was not truly completed until at least Lagrange's *Mechanique Analytique* of 1788 (or perhaps even the fusion of geometry and analysis constructed by Hamilton in the early nineteenth century.) The extension of equilibrium to states of systems in motion would be most explicitly and fully realized by Lagrange [1]. *Mechanique Analytique* is divided in two sections,

[1] [Lagrange (1788)]. D'Alembert earlier had remarked that he (d'Alembert) had "reduced dynamics to statics", through what we now know as d'Alembert's principle. But d'Alembert's approach was a methodology for solving problems in dynamics by imagining equilibrium as a counterfactual. [Alembert (1743/1921)]

$$S\left(\frac{d^2x}{dt^2}\delta x + \frac{d^2y}{dt^2}\delta y + \frac{d^2z}{dt^2}\delta z\right)m + S\left(P\delta p + Q\delta q + R\delta r + \ldots\right)m = 0$$

Fig. 4.1 Lagrange's Principle of Virtual Velocities. When the virtual velocity terms, δx, δy, etc. are zero, the equation gives the statics case

the first on statics, the second on dynamics. The principle equation of each section, the Principle of Virtual Velocities (see Fig. 4.1), is the same for each section, with the difference that the velocities are non-zero in the dynamics case. A novelty of Lagrange's presentation is that all the terms of the equation are written on the left-hand side and set equal to zero, thus making the equilibrium arguably more striking. Although the resulting equation "looks" less like a balance, it states more dramatically that a whole set of interactions and dynamic quantities literally comes to nothing. There are no figures in this work. The equations provide the only "mechanical" representations.

The conceptual advances of Euler, Lagrange and others in eighteenth-century mechanics were significant. Having an historiography which properly recognizes these developments is a great benefit. Still, we might worry that the unified story is a trivial one. We need to address, for example, the worry that if even Newton's non-contact, action-at-a-distance treatment of universal gravity amounts to a substantive appeal to the balance model, then most of the history of science will qualify as explanation through a balance, as explanation in terms of a simple machine. This, in turn, would imply that large swaths of the history of science could be construed as a part of the Mechanical Philosophy, threatening the usefulness of the concept to historiography. A closer examination of Machamer on Aristotle and natural motion is therefore in order. We will need to bear in mind the key points of the MMK account of the Mechanical Philosophy which connect equilibrium, natural motion, and mechanism, which are:

1. Equilibrium can include moving, changing systems.
2. The natural motions of the parts of a mechanism constrain one another, in ways which can be mathematically expressed (usually involving other physical features, such as mass or distance.)
3. The simple machines, especially the balance, are models for the mathematically expressed constraints.

4.3 Aristotle: Place and Rest, Equilibrium and Natural Motion

An early treatment by Machamer of natural motion can be found in "Aristotle on Natural Place and Natural Motion" ([Machamer (1978)]). Here Machamer offers an interpretation of the role of natural place in Aristotelian explanation of the natural motion of the elements. His response is a rejection of then standard attractionist readings of Aristotle whereby, for example, the center of the cosmos literally *pulls* earth toward it. Such readings are hard to reconcile with Aristotle's own claims that place, as a mere mathematical abstraction, cannot have causal efficacy.

Machamer, instead, understands Aristotle to be appealing to two senses of place: one is the mathematical abstraction, abstracted from the order of the cosmos (which Machamer calls the cosmological sense [Machamer (1978), 381]); the other is the de facto resting place of earth, defined by the innermost boundary of what surrounds (or would surround it were all the earth and only the earth located as close to the center of the cosmos as possible, surrounded by water, etc.) Only the latter sense, which Machamer calls the relative definition of place (Ibid), "presupposes the existence of a body which is contained by the boundary." (Ibid.)

Both senses of place play a role in Machamer's explication of the Aristotelian formal and final causes of natural motion.

> The formal cause is given in the first sense by the natural or organic unity that an element attains when in its natural place, in the second by the natural order of the cosmos.…In both senses of formal cause these are statements concerning the actualizing of potentials inherent in the essence of the bodies spoken about, that is, the elements or the cosmos. The final cause is given as the fulfillment of the form each element potentially has and by the fulfillment of all parts on the cosmological scale [Machamer (1978), 386].

The other function that natural place plays in explaining natural motion is that each natural place has the power of conferring rest on its respective element: earth rests at the natural place of earth, where it is heavy; water rests at the natural place of water, above earth and below air, where water is heavy insofar as the boundary of its place is defined by the air above, and light insofar as the boundary of its place is defined by the earth below.

This is what is meant by the fulfillment of the form of the element upon reaching its natural place.

> [W]hen Aristotle equates the natural with what is ordered…we can interpret this as saying that when things are ordered properly, they are as they should be by nature. Cosmologically speaking, this would be when all things are in their natural places…[Machamer (1978), 380]

There is no mention of equilibrium by Machamer here, and indeed the only role for actual rest in the explanation is a counterfactual one. There is a mention of mechanism, but only in the negative: "To my knowledge Aristotle is never very clear on the mechanism by which the sun moves or causes anything." (op cit, 383) The motion of the sun across the sky is the "general efficient cause of all coming-to-be or actualization" of elements, as the cause of their substantial transformations through heating or cooling.

Does Aristotle's explanation of natural motion (at least as Machamer understood it in 1978) qualify as a mechanism? Is Aristotle a part of "the" Mechanical Philosophy? Any adequate Aristotelian theory of machines will still have to save the phenomena. The phenomena of the lever or of the balance is that, in certain configurations, a less heavy object can support or even lift a heavier one. The ratios of mechanical advantage will apply. So within the broader Aristotelian theory of natural motion there will be a theory of machines which, on its own, could look like a Mechanical Philosophy — it would satisfy 2 and 3 of the MMK criteria. But it could not make sense within that theory of machines to think of their dynamic relations as a kind of equilibrium (criteira 1.) In the fuller (i.e. cosmological) explanation of Aristotelian natural motion, motion occurs as a striving toward rest, toward the equilibrium which is the state of a natural element in its natural place. In Mechanical Philosophy, rest is a special case of equilibrium; for Aristotle, one is the cause of the other. Even Guido Ubaldo, a representative of Renaissance mechanics, thought of mechanics as working against natural motion, "in rivalry with the laws of nature."([Ubaldo (1969), 241]. Quoted in MMK at [Machamer et al. (2012), 375].) The incompatibility is true of Aristotelian natural motion, but not of those which would come after. So the generality of the Mechanical Philosophy and of equilibrium on the MMK conception can still discriminate Mechanical Philosophy from Aristotelianism. I now turn to the role of equilibrium in the models of intelligibility described by Machamer and Woody (1994). The simple machines on which the models are based have equilibrium conditions — conditions which are relations among natural motions. Grounding the understanding provided by the models in natural motions allows us to avoid making too-strong ontological commitments.

4.4 Galileo: Models of Intelligibility

Machamer was first to emphasize the theme of equilibrium in the work of Galileo and to point out Galileo's consistent use of the balance as a model. Later, he and Andrea Woody, in a co-authored paper, would interpret Galileo's use of equilibrium and the balance model as a strategy for rendering natural phenomena intelligible. This interpretation, they would suggest, could provide insight into the teaching of science, the historiography of science, and even its advance.

The importance of equilibrium to the intelligibility of Galileo's models is explicitly highlighted by them.

> [F]or Galileo the whole schema of intelligibility becomes an equilibrium problem: What is the cause of (or force that causes) something becoming unbalanced? And what force will cause it to come back into balance? Where is the balance point? [Machamer and Woody (1994), 217]

There is more than one explanatory strategy implicit in this. First is the appeal to equilibrium itself. The conditions under which a machine (e.g. balance, lever, capstan, etc.) is at equilibrium are taken to be "easily comprehended" (p. 222). Deviations from equilibrium are another way of explaining phenomena. Deviations, too, it turns out, are in a kind of equilibrium according to Machamer and Woody. (I distinguish the two kinds of equilibrium as static and dynamic.) It is this aspect of their models of intelligibility I want to focus on.

As pointed out in the previous section, for Machamer a key advantage of explanations which appeal to equilibrium is that they can be extended to dynamic, changing states. Using the balance model allowed Galileo to do just that. Galileo used circles to geometrically represent the balance and the mathematical relationships among the parts of the machine. Weights on the arms of a balance would move on circles, the arms of the balance being the radii of the circles and the fulcrum as center. Galileo used the name moment (momento) for the combined ratio of the motion multiplied by the radius of the circle of that motion (i.e. the distance from the fulcrum in the case of the lever or balance.) A machine lets you move a larger weight with a lesser one: some parts will move farther or faster, some slower, but the moments will be in equilibrium. In the case of a lever, for example, even when the arms are in motion and the machine is in operation, we can still think of an equilibrium between the moments at each end of the lever, which is how the lesser weight can move the greater. "Galileo used the balance model to make intelligible problems of dynamics as well as kinematics" [Machamer and Woody (1994), 215].

Importantly, equilibrium of moments fits with Galileo's own conception of natural motion as circular. Machamer and Woody (MW) unpack Galileo's use of equilibrium and the balance model to illustrate their notion of a model of intelligibility. By grounding the models in natural motion we can further unpack the models, and the intelligibility of the models.

Models of intelligibility, according to MW, "direct" our thoughts in solving problems to the properties which are important [Machamer and Woody (1994), 219]. Properties which stand in equilibrium relations of the mechanism are picked out by their privileged role in the mechanism. There is coherence between the models and a broader scientific view when the privileged properties are the natural motions of bodies — i.e. the non-violent motions of matter, the characteristic behaviours of entities, or the definitive property (or properties) of matter. Some of these I've described above, and they may be up or down, uniform or accelerated, driven by an innate force or a state of motion; the measures of natural motions might be distance or force, or velocity or virtual velocity, or action or work, among many others. Natural motions and equilibrium have something in common where explanations are concerned: they are aspects of the phenomena which do not require explanation, but which can be used in constructing explanations. They are unexplained explainers. (I'm here drawing, as a reading of MW, a rough identity between intelligibility, being understood, and not requiring an explanation.)

Investigating the equilibrium conditions of machines such as the balance, the lever, or the pendulum, therefore provides problem-solving advantage through the simplification of problems, particularly by revealing some aspects of the phenomenon or phenomena as not needing an explanation, and also by providing a grounding of intelligible behaviours and properties out of which to build explanations. But the search for equilibrium conditions has also driven the historical development of natural philosophy.

> [W]e would suggest that historically without the intelligibility and form of understanding provided by the balance, and its mathematical tractability and experimental possibilities, the abstract concepts of the mechanical world picture would not have been either possible or plausible [Machamer and Woody (1994), 220].

Thinking about equilibrium has historical, philosophical, and practical importance.

Finally, by picking out privileged properties, models of intelligibility also play a normative role. The important properties are also the ones we ought to pay attention to. There are three ways to unpack the force of this 'ought' (cf. p224). The first sense of normativity derives from the problem-solving advantages provided by looking for equilibrium and balance relations. We ought to, in other words, use those models which contribute to our abilities to solve problems (since solving problems is a desirable outcome.)

The second derives from the historiographical insight we gain. Machamer's attention to equilibrium and the balance model have allowed him to give a unified, long and broad view of the developments of science. Just as developments in the models themselves contributed to revolutions, and the result of the revolutions was improved adequacy of our understanding of nature, we have an improved understanding of the regular, reliable processes behind scientific change by recognizing the role of these models. Models of intelligibility not only help to make natural phenomena intelligible, they help to make the history of understanding natural phenomena more intelligible.

There is a third, but potentially more problematic, way of understanding the normativity of the equilibrium-balance model: the mechanistic descriptions are prescriptive because they are true of the world. That is, the mechanistic explanations work because they are true, and so we ought to be looking for mechanisms. But we've been considering how widely applicable the mechanism approach is once we make balance or equilibrium among natural motions the defining feature. This approach therefore faces the problem of multiplying ontologies or devolving into triviality — mechanisms are everywhere and so too, therefore, are ontologies. I take up this worry in the final section.

4.5 Conclusion: History and Philosophy of Science, and Machamerical Explanation

Before we consider pluralism and mechanism, let's remind ourselves of the details and advantages which speak in favour of the account as now understood. Discovering equilibrium relations improves our understanding and problem solving abilities regarding the world by simplifying our pictures and rendering phenomena intelligible by analogy to simple machines such as the lever or balance. The concept of equilibrium can be extended to include dynamic systems, i.e. those undergoing changes but where those changes are regular and relate to one another through balanced exchanges. Balance and equilibrium among natural motions, as an historiographic concept, also allows us to tell long histories of scientific change, where many figures and episodes can be brought under a tradition of Mechanical Philosophy, from Archimedes through the Scientific Revolution (including the eighteenth-century) up to seemingly all the modern disciplines of science: an evolution in our descriptions of the world in terms of its entities and how they behave and interact by nature.

Equilibrium relations of the mechanical models hold among the natural motions of the parts of the mechanism. We have just described two senses in which these models can be normative according to Machamer. There is a third sense of normativity, according to which the mechanical models are best, or preferable, because the natural motions on which they are based are, in fact, true — true because they describe motions which belong to the real nature of real entities. The models are not just a convenient way of thinking about the world, they are a metaphysically incisive one."[M]odels of intelligibility...depart from this psychological class of concepts by existing, in some sense, apart from the person. They are not just abstractions, but can be instantiated by real concrete objects" [Machamer and Woody (1994), 221–222].

The problem with a realist interpretation of mechanisms is that it allows for only an uncomfortable scientific pluralism, where ontologies may be multiplied ad infinitum, anywhere a useful (e.g. stable, intelligible, expedient) input-output, balance relation can be posited. I suspect that Machamer, as a proponent of the mechanism account of explanation, might not be too concerned about this outcome. However, his strong (and I think wholly justified) belief in the value of the balance as the underlying model of intelligibility for science does not necessitate an equally strong commitment to a naive entity realism regarding the model. Moreover, according to Machamer's well known dualism, mechanisms consist of entities and activities [Machamer et al. (2000)]. This dualism diffuses some of the force of the objection. On Machamer's mechanistic picture we are not carving the world at its joints. We are parsing the behaviour of the world according to various dualistic pictures, even while the pictures are constrained by the way the world really is. These pictures can be constrained because they are mechanistic and so consist in relations among the parts of the picture, and those relations have physical instantiations. "In their physical presence actions can be performed on [the mechanical models] and relations can be literally discovered and seen among their parts, thus they lend themselves to experimental possibilities" [Machamer and Woody (1994), 222].

There is still room for realism, therefore, and for avoiding outright nominalism. One can replace Gabbey's historiographical distinction between mechanics and mechanical with a new distinction between machines and mechanisms. A mechanism is any one of the models of intelligibility known as the simple machines, or it is some combination of those. The operation of many physical systems can be understood through these models. We can call those physical systems machines, qua instantiations of a mechanism.[2] A machine, in other words, is some physical system whose operation can be explained through a mechanism.

[2]Machamer seems to endorse the distinction between mechanisms and machines. Regarding Galileo's use of the balance model he remarks "What made the model all the more satisfying was that the balance was an actual machine, which clearly instantiated the geometrical representations that he used to solve his problems." [Machamer and Woody (1994), 219]

What ties equilibrium, explanation, natural motion, and now pluralism together can be seen by recognizing what each discipline takes for granted in explaining phenomena. As I remarked in the beginning, calling a motion (or some change or activity) a natural motion is a way of bracketing off the need for explanation in certain regards. Appeals to equilibrium have long been a similar sort of explanatory strategy: a self-evident activity of entities which requires no explanation. Which equilibrium arrangements exist between which entities and in which circumstances are discipline specific. Discovering mechanisms is not the same as discovering ontology. Parsing the world in terms of networks of entities and activities should be understood as ontically neutral, even while allowing that such parsing can dramatically alter the affordances offered by a world parsed in those terms. Different disciplines will provide better explanations of some parts of the world than others because of the entities they refer to in describing the world and the motions which require no explanation because they are natural to those entities.

The discipline generic idea that entities might be engaged with one another through change-producing activities does not sound inherently mechanical. There are disciplines where mechanism might be an appropriate metaphor but machine is not. Rather than mechanical then, let's call dualistic explanations which appeal to a stock network of discipline-relevant unexplained explainers by a new name: Machamerical explanation. Machamerical explanation is a strategy for the scientist grappling with a complex phenomenon or system. It is also a strategy for the historian and philosopher of science struggling to find a unified objective picture behind the complexities and contingencies of the history of science — what Machamer called, in passing, in 1975 "the problem of the dynamics of change"[Machamer (1975), 374]. Pay attention, first, to the details of your subject. Look for what doesn't change, but not just fixed entities; look for patterns of entities engaged in regular interactions and exchanges.

> One becomes more philosophical when, after all this is done, one compares and contrasts two or more such cases and their meta-characterizations. The attempt, at this level, is to isolate the general characteristics which have persevered over time and which can be seen in various domains. Also, at this time, one brings in various explanatory hypotheses concerning general features of the rationale underlying changes [Machamer (1975), 380].

References

d'Alembert, Jean Le Rond. 1743/1921. *Trait'e de dynamique*. Paris: Gauthier-Villars.

Gabbey, Alan. 2002. Newton, Active Powers, and the Mechanical Philosophy. In *The Cambridge Companion to Newton*, ed. I. Bernard Cohen and George E. Smith, 329–357. Cambridge: Cambridge University Press.

Janiak, Andrew. 2008. *Newton as Philosopher*. Cambridge: Cambridge Univesity Press.

Lagrange, Joseph-Louis. 1788. *Méchanique Analitique*. Paris: Desaint.

Machamer, Peter. 1975. Understanding Scientific Change. *Studies in History and Philosophy of Science* 5: 373–381.

Machamer, Peter K. 1978. Aristotle on Natural Place and Natural Motion. *Isis* 69(3): 377–387.

Machamer, Peter, and J.E. McGuire. 2009. *Descartes's Changing Mind*. Princeton: Princeton University Press.
Machamer, Peter, and Andrea Woody. 1994. A Model of Intelligibility in Science: Using Galileo's Balance as a Model for Understanding the Motion of Bodies. *Science & Education* 3: 215–244.
Machamer, Peter, Lindley Darden, Carl Craver, et al. 2000. Thinking About Mechanisms. *Philosophy of Science* 67: 1–25.
Machamer, Peter, J.E. McGuire, Hylarie Kochiras, et al. 2012. Newton and the Mechanicalphilosophy: Gravitation as the Balance of the Heavens. *The Southern Journal of Philosophy* 50(3): 370–388.
Ubaldo, Guido. 1969. Selections from the Books of Mechanics. In *Mechanics in Sixteenth-Century Italy: Selections from Tartaglia, Benedetti, Guido Ubaldo, and Galileo*, ed. Stillman Drake and I.E. Drabkin. Madison: University of Wisconsin Press.

Chapter 5
Diagnosing Superstition: Superstition and Piety in Spinoza's Political Philosophy

Francesca di Poppa

Abstract The notion of superstition has a long history of being understood in terms of epistemic and psychological features, although many discussions include its problematic political consequences. I argue that Spinoza's discussion of superstition in *Theological-Political Treatise* is an exception. Spinoza connects superstition and piety with the problem of political stability via the notion of obedience, and uses the term "superstitious" to label religious attitudes and practices that undermine civil obedience by establishing demands of allegiance, on the part of the religious authority, that compete with those of the government. Contrary to existing interpretations, I will show that, for Spinoza, "superstition" is not intrinsically characterized by features such as intolerance and anti-intellectualism. In *Theological-Political Treatise*, practices that clearly shows these traits are labeled as pious, rather than superstitious, precisely because, within certain societies, they foster obedience and therefore stability. One of Spinoza's goals in TTP is to show that what counts as pious among the ancient Jews or in the Ottoman Empire should be considered superstitious in a more modern, diverse society such as the seventeenth century Dutch Republic, and that libertas philosophandi is the best way to inoculate a society (if not every single individual) against superstition.

Superstition is usually defined in terms of epistemic or psychological features. While it is acknowledged that superstitious beliefs and practices have important political consequences, the nature of superstition *per se* is defined independently of its political implications. This is also how most commentators have read the concept of superstition in Spinoza's *Theological Political Treatise (*TTP). Only a minority have interpreted "superstition" as a political concept, explaining it as a tool of absolutist rulers. In what follows, I will argue that superstition in TTP is, in fact, a political concept, but that it is not the tool of absolutist rulers; rather, the presence of

F. di Poppa (✉)
Texas Tech University, Lubbock, TX 79409, USA
e-mail: francesca.di-poppa@ttu.edu

M.P. Adams et al. (eds.), *Eppur si muove: Doing History and Philosophy of Science with Peter Machamer*, The Western Ontario Series in Philosophy of Science 81, DOI 10.1007/978-3-319-52768-0_5

superstitious attitudes undermines civil obedience and therefore produces instability, regardless of the form of government.

There is a continuity between Hobbes' and Spinoza's discussions of superstition, in the sense that for both superstition is a political notion focusing on the kind of religious attitude that challenges civil obedience. While there are important differences between Hobbes' and Spinoza's views on what is conducive to obedience, they both use the term "superstition" to stigmatize religious attitudes that create demands that can undermine civil obedience.

5.1 A Short Literature Review

Recent discussions of superstition usually emphasize two aspects: epistemic and psychological.[1] For example, Curley stresses that superstition consists of false beliefs about God, such as, for instance, providence and the afterlife.[2] Curley then distinguishes between "good" and "bad" superstition, depending on whether or not the attitudes and practices stemming from those beliefs are compatible with good citizenship. Curley singles out exclusivist beliefs in particular as most divisive: the conviction that only those who share certain beliefs and practices will gain salvation is conducive to sectarian hate and is therefore incompatible with a flourishing civil society. I find myself in agreement with Curley's emphasis on good vs. bad citizenship, but, as will become clear, his examples are problematic. Also, Curley's interpretation does not correspond to Spinoza's use of the terms "piety" and "superstition," as we will see.

Levene, in *Spinoza's Revelation: Religion, Democracy, and Reason*, Ch. 1, also focuses on false religious beliefs stemming from fear and hope. In her view, superstition includes, in addition to false beliefs about God, an attitude that she sometimes calls "greed," but that goes beyond mere greed in its goal to seek self-aggrandizement at the expense of others.[3] So, Levene remarks, Spinoza's tirades against superstition should be understood as targeting self-serving theologians, rather than the masses, as responsible for the false belief in a transcendent God, which is intrinsically debasing for humanity.

James, in *Spinoza on Philosophy, Religion, and Politics*, argues that epistemic considerations are irrelevant to understanding of superstition as discussed in TTP, and that its main feature is a form of aggressive anti-intellectualism. It is this toxic

[1] There is no space here for a comprehensive literature review, but it should be pointed out that, in his *Spinoza's Critique of Religion*, Ch. IX, Leo Strauss also interprets "superstition" as connected to irrationality and hate.

[2] "Spinoza's Exchange with Albert Burgh," in Melamed and Rosenthal 2010. In a 1986 essay republished in 2011, "Politique et religion chez Hobbes et Spinoza," Alexandre Matheron similarly defines superstition in terms of false beliefs, in particular belief in revelation and in the necessity of specific forms of worship.

[3] See, for example, p. 17.

psychological trait, not false beliefs, that turns superstition into a political problem. James' remarks certainly apply to a large number of philosophical discussions of superstition (e.g. Francis Bacon), but, as I will show, they do not apply to Spinoza's. Rosenthal offers a somewhat similar reading, arguing that superstition consists of false theological tenets unaccompanied by justice and charity.[4]

All these interpreters agree on the reading that the very real distinction between superstition and piety in TTP is based on psychological or epistemic features. I will argue that these readings fail to capture the nature of the distinction, which is based on political considerations.

Interpretations of superstition that stress its political nature exist (e.g. Smith, Tosel),[5] but they define superstition as an instrument of absolute power. Just as the psychological/epistemic readings, they are problematic in the light of the texts that I will discuss in a later section.

5.2 Superstition in TTP

The term "superstition" in TTP is introduced in the very first sentence. Here, "superstition" largely overlaps with the discussion in *Ethics*: a damaging religious attitude triggered and fueled by the 'evil twins' hope and fear, and that in turn feeds them in a vicious circle.

Soon enough, however, it becomes clear that the notion of superstition in TTP diverges from that of *Ethics*. First of all, in TTP "superstition" is not primarily opposed to what is called in *Ethics* "true religion," but to "piety" or "faith." What makes the tenets listed in TTP Ch. 14 tenets of piety, or faith, is not their being true (as I will show, some are clearly incompatible with true, philosophical religion), but their being conducive to *civil obedience*.

Superstition in the Preface of TTP still seems to be opposed to piety in the sense that, while neither is a form of *vera religio*, they are accompanied by very different psychological attitudes: piety involves charity and love, while superstition involves hatred and fear of those who believe differently. This turns the pious into model citizens and the superstitious into nuisances at best.

The problem with defining piety and superstition in terms of such psychological features, however, is that we can find egregious examples of attitudes that ought to be labeled as superstitious, according to this criterion, but which Spinoza calls pious. One example is found in Spinoza's account of the Hebrew state in Ch. 17: practices that involve rigid obedience to rules, and a certain amount of 'hate speech'

---

[4] "Miracles, wonder, and the state in Spinoza's *Theological-Political Treatise*," also in Melamed and Rosenthal (eds.), 2010.

[5] Smith, *Spinoza, Liberalism, and the Question of Jewish Identity*, and Tosel, "Superstition and Reading" (in Montag and Stolze, eds. *The New Spinoza*, 146–167), consider superstition a tool used by absolute rulers to foster unquestioning obedience.

against other nations, are praised as pious.[6] These attitudes and practices have all the marks of what the scholars I have briefly cited define as "superstitious," yet Spinoza fails to label them as such. The negative term "superstition" is only introduced when discussing the priests' *coup* that causes the dissolution of the Hebrew state, as I will show soon.

Considering that the distinction between superstition and piety is clearly important to Spinoza, then, it seems worthwhile to find a better interpretation. I will argue that Spinoza intended the distinction between piety and superstition to be based on whether or not religious attitudes and practices foster civil obedience, and therefore stability, or rather pose problems of competing allegiance,[7] and that this discussion has much in common with Hobbes' discussion in *Leviathan*.

5.3 Superstition in Hobbes' *Leviathan*

Thomas Hobbes' *Leviathan* was published in English in 1651. By 1668, Hobbes' Latin translation[8] and a 1667 Dutch translation by Abraham van Berkel were circulating in the Dutch Republic.[9] Spinoza was familiar with Hobbes' philosophy. He openly discussed and criticized it, and there is evidence that Hobbes' ideas exercised some influence on Spinoza, directly or through contemporary discussions.[10]

---

[6] The term "pious" is a positive term in TTP, because piety and faith, as Spinoza makes explicit in Ch. 14, are conducive to obedience and therefore to a peaceful and thriving state. I take it to be self-evident that "superstitious" is a derogatory term.

[7] *Contra* Smith and Tosel.

[8] An edition in 8 volumes of Hobbes' works in Latin, including *Leviathan*, was published in 1668 by Johan Blaeu in collaboration with Hobbes. Hobbes' work generated much discussion, in particular in the works of the brothers Pieter and Johan de la Court. Between 1660 and 1663 Pieter de la Court (Johan died in 1660) published two books and several pamphlets on political theory, largely indebted to Hobbes' discussion. See Blom, *Morality and Causality in Politics The Rise of Natural in Dutch 17th Century Political Thought,* Israel, *The Dutch Republic: Its Rise, Greatness, and Fall. 1477–1808* and *Radical Enlightenment: Philosophy and the Making of Modernity 1650–1750;* Wyger R.E. Welema, *Republicans. Essays on Eighteenth Century Dutch Political Thought.*

[9] Two important members of Spinoza's "circle of friends," Lodewijk Meyer and Adriaen Koerbag, knew van Berkel. See also Blom, Ch. 7 and 8; van Bunge, *From Stevin to Spinoza: An Essay on Philosophy in the 17th Century Dutch Republic,* Ch. 4; Frijhoff and Spies, *Dutch Culture in a European Perspective*: *1650, Hard-Won Unity*, Ch. 5 and 6; Nadler's *Spinoza. A Life*, Ch. 10; Israel 2002, Ch. 8 and 9, on the influence of Hobbes' political ideas on Spinoza's milieu.

[10] The extent of Hobbes' influence on Spinoza is a matter of debate. Hobbes' ideas were widely discussed, for example, in Franciscus van den Enden's house, which Spinoza attended in the mid '50s for Latin lessons. Van den Enden's influence on the young Spinoza has been overestimated by Wim Klever in his 1991, but Nadler, among others, argues that it was in van den Enden's house that Spinoza was exposed to Bruno, Machiavelli, Hobbes, Grotius, Tacitus, and other important political thought. Spinoza then became acquainted with Hobbes' political ideas long before the translation of *Leviathan* became available. I believe that a case can be made for a direct influence of the Hobbesian notion of superstition on Spinoza, but I cannot defend such claim in this paper.

In Ch. VI of *Leviathan*, dedicated to the passions, superstition is defined in a brief, controversial paragraph:

> *Fear* of power invisible, feigned by the mind, or imagined from tales publicly allowed, RELIGION; not allowed, SUPERSTITION. And when the power imagined is truly such as we imagine, TRUE RELIGION.[11]

This paragraph caused quite the stir, from early critics such as Alexander Ross' *Leviathan Drawn out with a Hook*[12] to A. P. Martinich's accusation of philosophical incompetence,[13] for its apparent claim that superstition and true religion can coincide. Curley finds this passage problematic precisely because of "the natural assumption that true religion is a species of religion."[14] An analogous problem is found in Hobbes' discussion of prophecy.[15]

I cannot do justice here to the *vexata quaestio* of Hobbes' genuine thoughts on religion. Fortunately, it is not necessary for my purposes. My interpretation is based on the claim that the definition of superstition in Ch. VI is clearly in *political* terms, regardless of whether or not it is consistent with the rest of his philosophy (I believe that it is), or consistently applied through his work (I believe that it is not).

Hobbes tells us that all religious attitudes are based on "fear of invisible powers." What distinguishes religion *in general* from superstition does not have anything to do with the epistemic features of the beliefs that this fear generates. The clause "publicly allowed" must refer to the *tales* that generate the fear, rather than to the fear itself: the government has no direct control over what a mind feigns or fears. So, superstition is the religious attitude that stems from tales forbidden by the government. Hobbes could have been clearer here, but he seems to be implying that there is no superstition in the private sphere. Superstition comes into being when individuals move from privately held beliefs to shared narratives and practices, which the sovereign can legitimately forbid.

---

[11] All quotes from *Leviathan* are from Thomas Hobbes: *Leviathan.* Edited with Introduction and Notes by Edwin Curley. (from now on, L followed by page number). L 31.

[12] *Leviathan drawn out with a hook, or, Animadversions upon Mr. Hobbs his Leviathan* by Alex. Rosse. London: Printed by Tho. Newcomb, for Richard Royston, 1653. On early criticism see Jon Parkin, "The Reception of Hobbes' Leviathan," in *The Cambridge Companion to Hobbes' Leviathan,* 445–459.

[13] A. P. Martinich, *Hobbes,* 198. Martinich, who points out that, in *De Cive* 16.1, Hobbes had offered an impeccable definition of "superstition" as *irrational* fear of invisible powers, suggests that, rather than reading this as a philosophically incompetent definition, we should understand Hobbes as somewhat ironically referring to the common, and unphilosophical, use of the term. I find this explanation somewhat *ad hoc*, since the supposedly ironic definition is placed in a list of definitions that Hobbes intended quite seriously. See also his *The Two Gods of Leviathan.*

[14] Curley, "'I Durst Not Write So Boldly' or, How to Read Hobbes's Theological-political Treatise." In Bostrenghi, ed., *Hobbes e Spinoza, Scienza e Politica.*

[15] See *Leviathan* XXXII on the criteria for prophecy: "One is the doing of miracles; the other is the not teaching any other religion than that which is already established." (L 247–8) By these standards, Christian prophecy becomes false prophecy wherever Christianity is forbidden. See Curley's and Martinich's discussions.

A similar point is made in the following passage from Ch. XI, where no reference is made to legal status:

> And this fear of things invisible is the natural seed of that which every one in himself calleth religion; and in them that worship or fear that power otherwise than they do, superstition.[16]

So, true religion and superstition can overlap.[17] The term "religion" applies to the fear of invisible powers *in general*. "Superstition" applies only to fear connected to views about such powers that are not allowed by the sovereign, who, in Hobbes' view, has both the duty and the right to decide which opinions, being compatible with the peace of the state, are to be allowed in the public sphere.[18] Those opinions deemed (rightly or wrongly) conducive to disobedience and instability, and therefore forbidden, are superstitious.

Yet, even when originated from illegal tales, this fear may well be of powers that are really as imagined. When this happens, it is *both* superstition *and* true religion. Hobbes was openly skeptical about the human ability to distinguish true from false revelations, even when they are experienced in person, and it is clear that for him, as Matheron explains, whether religious beliefs are true or false is politically irrelevant.[19] Christianity, which Hobbes alleges to hold as true,[20] was, at some point, illegal, and therefore, according to his definition, superstitious. Hobbes even addresses, in Ch. XLIII, the plight of a Christian in a state where her religion is forbidden, and therefore superstitious: the Christian subject has no right to disobey, but must keep her faith "invisible."[21]

A passage from Ch. III emphasizes that Hobbes' notion of superstition is connected with political instability, rather than with false religious beliefs. Here, Hobbes seems to acknowledge what is not quite explicit in his definition of superstition from Ch. VI: that there are religious beliefs, attitudes, and practices that

[16] L 63.

[17] This passage has a different tone from the passage from Book VI, because here Hobbes refers to how individuals label their religious views vs. their opponents'. However, it is consistent with my reading of the definition in *Leviathan* VI.

[18] Ch. 18: "... it is annexed to the sovereign to be judge of what opinions and doctrines are averse, and what conducing, to peace; and consequently, on what occasion, how far, and what men are to be trusted withal, in speaking to multitudes of people, and who shall examine the doctrines of all books... For... in the well-governing of opinions consisteth the well-governing of men's actions, in order to their peace and concord." L 113. One of the most striking differences between Hobbes and Spinoza is in their views of the extent to which freedom of expression should be granted.

[19] Matheron 2011.

[20] Martinich, in 1992, argues that Hobbes was an orthodox Calvinist. I find Curley's rebuttal (e.g. "Calvin and Hobbes, or, Hobbes as an Orthodox Christian") quite persuasive. Also, as Springborg writes in "Hobbes on Religion" (in Springborg, ed., 1996, 346–380), Hobbes' 'Calvinism' failed to impress the Low Countries Synods, who banned *Leviathan*.

[21] L 410.

undermine civil obedience even when, being allowed, engaging in them does not constitute actual disobedience.[22]

> ...there is no doubt but God can make unnatural apparitions: but that He does it so often as men need to fear such things more than they fear the stay, or change, of the course of Nature... is no point of Christian faith. ... If this *superstitious* fear of spirits were taken away, and with it prognostics from dreams, false prophecies, and many other things depending thereon, by which crafty ambitious persons abuse the simple people, *men would be would be much more fitted than they are for civil obedience* [my emphasis][23]

In what follows, I will argue that Spinoza's notion of superstition in TTP is a political notion in the same sense and in the same way in which Hobbes' is. Both authors are concerned with the connection between religious attitudes and political stability.

5.4 Superstition As Destabilizing Religious Attitude

The term "superstition" is introduced in the first sentence of the Preface to Spinoza's TTP:

> If men were able to exercise complete control over all their circumstances, or if continuous good fortune were always their lot, they would never be prey to superstition. But since they are often reduced to such straits as to be without any resources, and their immoderate greed for fortune's fickle favours often makes them the wretched victims of alternating hopes and fears, the result is that, for the most part, their credulity knows no bounds[24]

Spinoza opposes superstition to piety, promising to show that freedom of inquiry, far from endangering *pietas*, strengthens it, or, as Spinoza puts it, *pietas* (as opposed to superstition) *depends on* freedom of inquiry.[25]

It seems that, *prima facie*, the distinction between superstition and piety or religion is based on the former's being wholly irrational:

---

[22] Catholicism, for example, by establishing the Pope as a vicar for Christ on earth, is inherently destabilizing, even when legal and therefore not falling under the definition of "superstition" in Ch. VI. In Ch. XLIV, opening Part IV, "On the Kingdom of Darkness," Hobbes explains that the Pope' demands for allegiance from all Christians encourage disobedience.

[23] L 11.

[24] TTP 1.

[25] It is useful to note, at this point, that Spinoza's connection between *pietas* and *libertas philosophandi* (which I prefer to translate as "freedom of inquiry") is meant to apply to specific historical and political circumstances, rather than being a universal norm. As others have pointed out, Spinoza's arguments for freedom of inquiry and expression (including religious expression) are based on pragmatic considerations, rather than some notion of inalienable rights. For Spinoza, such freedom is recommended to the extent to which it is conducive to political stability, and what is conducive to stability in the 17th century Dutch Republic is different from what is conducive to stability in the Hebrew state under Moses. See, among many other discussions, J. Smith, "Spinoza's Curious Defense of Tolerance," in Melamed and Rosenthal, eds., 2010, Rosenthal, "Spinoza's Republican Argument for Toleration" and "Spinoza on Why the Sovereign Can Command Men's Tongues But Not their Minds" (in Williams and Waldron, eds., 2008, 54–77), and Israel 2002.

> Piety and religion... take the form of ridiculous mysteries, and men who utterly despise reason, who reject and run away from the intellect as naturally corrupt--these are the men ... who are believed to possess the divine light... When I pondered over these facts, that the light of reason is not only despised but condemned by many as a source of impiety, that merely human suppositions are regarded as divine doctrines and that credulity is looked upon as faith... I deliberately resolved to examine Scriptures afresh, conscientiously and freely...

In these passages from the Preface, the three listed ingredients of superstition are fear, lack of rationality (or even hostility to reason), and obstinate dogmatism. The passage seems to imply that the last two do not belong to piety at all. The notion of superstition in the Preface seems to largely overlap with that found in *Ethics;* these passages largely corroborate, for example, James' 2012 discussion of superstition.

As we move along, though, we find examples of religious attitudes and practices that show all the ingredients listed above; yet, Spinoza labels them as *pious*, rather than superstitious. In Ch. 17 and 18, Spinoza discusses Moses' Hebrew state and its later disintegration. Primitive, less educated populations, he argues, are easily persuaded to obedience by religious narratives and practices that will not work with more sophisticated populations.[26] Moses' theocracy equated allegiance to the ruler with allegiance to God, and eliminated the distinction between civil obedience and worship. Worship, however, included practicing hateful invective against other nations and peoples:

> Therefore the patriotism of the Hebrews was not simply patriotism, but *piety*, and this, together with hatred for other nations, was so fostered and nourished by their daily rituals that it inevitably became part of their nature... Hence this daily invective… was bound to engender a lasting hatred of a most deep-rooted kind, since it was a hatred that had its source in strong devotion of *piety*... [my emphasis][27]

In this chapter, Spinoza endorses ritualistic expressions of hate towards other nations, as well as unquestioning obedience to norms minutely regulating daily life, as forms of *piety*, rather than superstition. Yet, these attitudes present precisely the kind of epistemic and psychological aspects associated with superstition historically, as well as in most readings of Spinoza's philosophy.

The reason for the label "pious" or "religious," rather than "superstitious," lies in the relationship between these attitudes and political stability. Mosaic practices created stability by promoting obedience to the ruler and creating a stronger sense of community among the members of God's elected people. Obedience to the ruler was willingly given because Moses persuaded his people that it was obedience to God himself. Spinoza describes a successful absolute theocracy that stopped being

[26] Spinoza's discussion of politics in TTP is a balancing act between his clearly expressed preference for democracy and the conclusion, supported by a series of historical examples, that each community has traits and constraints that might not make it suited for the most desirable form of government. See James, "Democracy and the good life in Spinoza's philosophy," in Huenemann, ed., 2008, 128–146.

[27] TTP 198. Shirley argues that Spinoza slanders his former correligionists. Shirley may well be correct about Spinoza's intentional misrepresentations; at the very least, he lacks charity and nuance.

successful precisely when competing claims to obedience were allowed to take root by giving power to the priestly caste of the Levites. In Ch. 18, Spinoza explains how this caused the decline of the Jewish state:

> There were no religious divisions among the people until the high priests in the second state acquired the authority to issue decrees and to transact government business-an authority which they sought to render permanent by usurping the government... They extended pontifical authority to all areas... As a result, *religion degenerated into pernicious superstition* [my emphasis].[28]

Spinoza intends to draw some general conclusions from his discussion of the Hebrews. His conclusion is not that the practices that worked for ancient Hebrew tribes will work for the Dutch Republic, notwithstanding the Dutch's fondness for comparing themselves with the Jews for having escaped from captivity.[29] The conclusion is, rather, that a ruler must be aware of what kind of practices will foster or undermine obedience among the non philosophically minded *of her own state.*

What was conducive to cohesion and flourishing for a people who had only known slavery for a long time cannot easily be transferred to very different political circumstances. The hate speech described in Ch 17 would have a very different effect when practiced by a group within a diverse nation. The attitude becomes then superstitious, *regardless of whether or not it is allowed*, because, by threatening the spirit of citizenship, it can undermine civil obedience. Clearly, Spinoza wants his readers to think about the intolerant attitude of the Dutch Counter-Remostrants and their attempt to use political power to impose their orthodoxy.

The split between the Remostrant and the more conservative Counter-Remostrant (Reformed) churches, following the 1618–19 Synod of Dort, was caused, among many other factors, by a very different understanding of the relationship between civil and religious authority. The Remostrants ascribed the church purely spiritual duties, and denied it the civil authority to censor and prosecute dissent. They accepted that most religious attitudes, insofar as they are internal matters of faith, are compatible with good citizenship, which resulted in an implicit endorsement of limited religious tolerance. The Reformed church, which was also virulently opposed to the teaching of Cartesianism in universities, wanted the government to be subject to church authority on matters of religion, including policies on censorship and prosecution of heresy.[30] This power struggle between church and state, and the aggressive interference of the Reformed church in government affairs, besides being generally destabilizing, threatened several members of Spinoza's community.

---

[28] TTP 208.

[29] On the multiple aspects of the narrative comparing 17th century Holland to the Old Testament Jews (from the liberation from Spanish domination to the reclamation of the land after a flood), see Schama, *The Embarrassment of Riches: An Interpretation of Dutch Culture in the Golden Age,* especially Ch. 2.

[30] See Israel 1995 Ch. 20 and 34, and 2002; Prak, *The Dutch Republic in the Seventeenth Century: The Golden Age.*

Among their victims was Spinoza's friend Adriaen Koerbagh, who was tried, sentenced, and imprisoned for his writings, and died in an Amsterdam prison in 1669.[31]

It becomes clear, then, that Spinoza, who endorses a universal faith as an instrument of public unity and peace, considers the Reformed church superstitious because of its attitude towards the civil power, not because of especially unenlightened tenets of faith. That this is the core of his concept of superstition is shown by another revealing passage in the Preface, where Spinoza uses the Ottoman Empire as the ultimate example of a repressive theocracy. Spinoza goes so far as to call Ottoman religious practices a *remedy* against the tendencies that create superstition, although these practices present (in his discussion) the dogmatism and aggressive anti-intellectualism associated with it:

> To counteract this *unfortunate tendency*, immense efforts have been made to invest *religion*, true or false, with such pomp and ceremony that it can sustain any shock and constantly evoke the deepest reverence in all its worshippers. In this the Turks have achieved the greatest measure of success.[32] They hold even discussion of *religion* to be sinful, and with their mass of dogma they gain such a through hold on the individual's judgment that they leave no room in the mind for the exercise of reason... [my emphasis][33]

The tendency referred to is the fluctuation of fear and hope, which draws humans to whatever is new and is not yet known not to work. This tendency cannot be fully eradicated; yet, it needs to be squelched for the sake of political stability. While manipulation and aggressive censorship are not universally desirable, they are legitimate tools for the promotion of civil obedience and the prevention of unrest.

This is why the Turks' beliefs and practices are labeled as "religion," rather than "superstition." Just as Moses' Hebrews, they (at least in Spinoza's discussion of them) are characterized as aggressively anti-intellectual and intolerant; yet, they are given the positive (in this context) labels of "religious" or "pious." Ottoman religious leaders use the trappings of pomp and mysteries, as well as a crushing censorship, to fight the destabilizing influence of the passions of hope and fear on the immature, uneducated minds of the populace. Moses built a theocracy where every minute aspect of daily life was dictated by religious law. Both are successful at maintaining stability: hence, these practices and attitudes are considered pious.

Spinoza's claim that such practices can contribute to a thriving state seems counterintuitive. In order to understand it, we need to remember that, in *Ethics,* Spinoza argues that humans tend to resist disempowering views, and to embrace those that make them feel in control, no matter how irrational such views might be.[34] Even

---

[31] Israel 2002, Ch. 10.

[32] It is unclear how familiar Spinoza was with the Ottoman Empire, which, while beginning its decline from the splendors of the 15th and 16th centuries, was still relatively tolerant and culturally lively. In Ch. 20, Spinoza notes that a theocracy such as the Hebrew state would be impossible in the 17th century, unless it renounced any outside trade. But he probably knew that the Ottomans engaged in flourishing trade with many states. So, his view of Ottoman Turkey does not stand to scrutiny. It seems that, just as with the Jews, Spinoza is picking a "low hanging fruit," a target that can be the object of harsh criticism without backlash.

[33] TTP 2–3.

[34] *Ethics* III, in particular P 12 to 15 and P 50, Schol.

philosophers, while striving for rationality, lead a large part of their life guided by their imagination.[35] Non-philosophers will be continuously enthralled by their imaginary narratives, having no access to the kind of adequate ideas that will rule out such narratives. Rather than contemptuously criticizing such tendencies, Spinoza remarks that we need to look at such narratives as survival strategies: they need to be evaluated based on whether or not they are conducive to survival and prosperity. While Spinoza does not think very positively of the kind of political stability that turns citizens into puppets or slaves (as he tells us in Ch. 20), it is still more desirable than a state of civil unrest.[36]

The Jews' narrative of being elected among the nations is empowering, both in times of success and in times of adversity. When a whole nation is convinced that God is on their side, rewarding loyalty and punishing faithlessness, even hate speech can be conducive to a sense of community. Things change dramatically when this "us vs. them" view is held by a religious community within a religiously diverse nation.[37] Analogously, religious rituals, dogmas, and a strict censorship can be seen as empowering to the extent that they bring the faithful to commit fully to their religious as well as political community; but, just like the rituals among Jews expressing hate for gentiles, they become politically problematic in societies with a certain amount of diversity. More importantly, for the purposes of Spinoza's discussion, religious intolerance and censorship become downright dangerous when imposed on societies used to a certain amount of freedom, such as the Dutch Republic.

We see, then, that the use of the terms "superstition" and "piety" in TTP is connected to the political notions of obedience and stability, rather than to epistemic or psychological traits. This is made quite explicit when Spinoza defines piety in terms of obedience:

> A man's beliefs should be regarded as pious or impious *only insofar as he is thereby induced to obey the moral law*... Therefore, if anyone by believing what is true becomes self-willed, he has a faith which in reality is impious; and if by believing what is false he becomes obedient ... he has a faith that is pious.[38]

While there is no space here for a detailed examination of the "tenets of universal faith" in Ch. 14, it is clear that, being based on the emotions of hope and fear (the same emotions that are described as causes of superstition in the Preface), they are

[35] The philosopher's 'bondage' is nicely discussed in James 2012, Ch. 8.

[36] In *Political Treatise* VI, Spinoza expresses a dismal view of the repressive, yet stable Ottoman Empire. See Curley, "Kissinger, Spinoza, and Genghis Khan," in Nadler, ed., *The Cambridge Companion to Spinoza*, 315–342.

[37] Once again, what is said about the Jews needs to be read with an understanding that Spinoza had contemporary events in mind, especially the attitude of the Counter-Remostrant churches engaged in a similar inflammatory, "us vs. them" rhetoric. Just as for Moses' Jews, such rhetoric might have worked towards fostering political cohesion in a religiously unified state, but the Dutch Republic was a very different political reality.

[38] TTP 157.

pious, rather than superstitious, to the extent that they are conducive to obedience and therefore stability.[39]

> ... Faith demands piety rather than truth; faith is pious and saving only by reason of the obedience it inspires, and consequently nobody is faithful except by reason of his obedience.[40]

The tenets, Spinoza writes, are vague enough that anyone who is not a philosopher should be allowed to interpret them as they please, as long as the interpretation is conducive to obedience, and no one should be allowed to aggressively impose their own reading on others. This is why Spinoza concludes that piety depends on a certain amount of freedom, in particular the freedom from the intrusion of the churches on personal matters of faith.[41] In discussing the Apostles in Ch. 11, Spinoza uses the term "superstition" to refer to proselytizing that, by injecting the rigors of (bad) philosophy in what should have been a simple message of obedience and charity, "gave rise to many disputes and schisms to vex the Church continually right from the time of the Apostles."[42]

5.5 Conclusion

The notion of superstition has a long history of being understood in terms of epistemic and psychological features, although many discussions include its treacherous political consequences. In this paper, I have argued that Spinoza's discussion of superstition in TTP is an exception. Spinoza connects superstition and piety with the problem of political stability via the notion of obedience, and uses the term "superstitious" to label religious attitudes and practices that promote civil disobedience by establishing demands of allegiance, on the part of the religious authority, that compete with those of the government. In making this philosophical move, Spinoza adopts a strategy that strongly resembles Hobbes', although the differences between Spinoza's and Hobbes' political philosophies should not be underestimated. Just like Hobbes, Spinoza targets contemporary views: in particular, the largely successful attempts by the Counter-Remostrants to use the government to impose its orthodoxy. Contrary to Hobbes, Spinoza does not explicitly define

[39] E.g.: "7. God forgives repentant sinners. There is no one who does not sin, so that, without this belief, all would despair of salvation, and there would be no reason to believe that God is merciful..." (TTP 126). For different reasons, James 2012 and Carlos Fraenkel, in "Spinoza on Miracles and the Truth of the Bible," argue that Spinoza considered these tenets flexible enough to be compatible with his philosophy. I find that at least three of them (including the one just quoted) are inconsistent with *Ethics*, and I do not believe that Spinoza would have applied them to the philosopher, who does not need piety in order to respect the law, and who is not, strictly speaking, obedient. Unfortunately, I cannot here address James' and Fraenkel's arguments in detail.

[40] TTP, 163.

[41] See Israel 2002, and his Introduction to the 2007 edition of TTP, viii–xxxiv.

[42] TTP, 144.

"superstition" in terms of what is or is not allowed, although, in his discussion of seditious speech in Ch. 20, he comes close to implying that superstition becomes seditious when it becomes too influential.

There are interesting implications to Spinoza's concept of superstition, and I cannot discuss them here. One of them is the implicit claim that, while it is impossible to 'vaccinate' each individual against superstition, *libertas philosophandi* goes a long way towards building enough 'herd immunity' that superstition will not become a threat. But the wise politician will be watchful.

References

Blom, Hans. 1995. *Morality and Causality in Politics The Rise of Natural in Dutch 17th Century Political Thought*. Utrech: Utrech University Press.

Bostrenghi, Daniela, ed. 1992. *Hobbes e Spinoza, Scienza e Politica*, 497–593. Naples: Bibliopolis.

Curley, Edwin. 1996. Calvin and Hobbes, or, Hobbes as an Orthodox Christian. *Journal of the History of Philosophy* 34(2): 257–271.

de Spinoza, Benedictus. 2001. *Theological-Political Treatise*. Translated by Samuel Shirley, Introduction by Seymour Feldman. Indianapolis: Hackett Publishing Company.

de Spinoza Benedictus. 2007. *Theological-Political Treatise*. Edited by Jonathan Israel. Cambridge: Cambridge University Press.

Fraenkel, Carlos. 2013. Spinoza on Miracles and the Truth of the Bible. *Journal of the History of Ideas* 74(4): 643–658.

Frijhoff, Willem, and Marjike Spies. *Dutch Culture in a European Perspective: 1650, Hard-Won Unity*. Vol. 2004. London: Palgrave MacMillan.

Hobbes, Thomas. 1994. *Leviathan. With selected variants from the Latin edition of 1668. Edited with Introduction and Notes by Edwin Curley*. Indianapolis: Hackett Publishing Company.

Huenemann, Charlie, ed. 2008. *Interpreting Spinoza: Critical Essays*. Cambridge: Cambridge University Press.

Israel, Jonathan. 1995. *The Dutch Republic: Its Rise, Greatness, and Fall. 1477–1808*. Oxford: Oxford University Press.

———. 2002. *Radical Enlightenment: Philosophy and the Making of Modernity 1650–1750*. Princeton: Princeton University Press.

James, Susan. 2012. *Spinoza on Philosophy, Religion, and Politics. The Theologico-Political Treatise*. Oxford: Oxford University Press.

Klever, Wim. 1991. A New Source of Spinozism: Franciscus van den Enden. *Journal of the History of Philosophy* 29: 613–631.

Levene, Nancy. 2009. *Spinoza's Revelation: Religion, Democracy, and Reason*. Cambridge: Cambridge University Press.

Martinich, A.P. 1992. *The Two Gods of Leviathan*. Cambridge: Cambridge University Press.

———. 2005. *Hobbes*. New York: Routledge.

Matheron, Alexandre. 2011. *Études sur Spinoza et les philosophies de l'âge classique*. Lyon: ENS Editions.

Melamed, Yitzhak, and Michael Rosenthal, ed. 2010. *Spinoza's Theological-Political Treatise. A Critical Guide*. Cambridge: Cambridge University Press.

Montag, Warren, and Ted Stolze, ed. 1997. *The New Spinoza*. Minneapolis: University of Minnesota Press.

Nadler, Steven, ed. 1996. *The Cambridge Companion to Spinoza*. Cambridge: Cambridge University Press.

———. 1999. *Spinoza. A Life*. Cambridge: Cambridge University Press.

Prak, Maarten. 2005. *The Dutch Republic in the Seventeenth Century: The Golden Age*. Cambridge: Cambridge University Press.
Rosenthal, Michael. 2003. Spinoza's Republican Argument for Toleration. *Journal of Political Philosophy* 11(3): 320–337.
Schama, Simon. 1988. *The Embarrassment of Riches: An Interpretation of Dutch Culture in the Golden Age*. Berkeley: University of California Press.
Springborg, Patricia. 2007. *The Cambridge Companion to Hobbes' Leviathan*. Cambridge: Cambridge University Press.
Steven, B. 1997. *Smith, Spinoza, Liberalism, and the Question of Jewish Identity*. New Haven: Yale University Press.
Strauss, Leo. 1963. *Spinoza's Critique of Religion*. New York: Schocken Books.
van Bunge, Wiep. 2001. *From Stevin to Spinoza: An Essay on Philosophy in the 17th Century Dutch Republic* Leiden: Brill.
Welema, Wyger R.E. 2007. *Republicans. Essays on Eighteenth Century Dutch Political Thought*. Leiden: Brill.
Williams, Melissa S., and Jeremy Waldron, ed. 2008. *Toleration and Its Limits*. New York: New York University Press.

Chapter 6
Science, Values, and Citizens

Heather Douglas

Abstract Science is one of the most important forces in contemporary society. The most reliable source of knowledge about the world, science shapes the technological possibilities before us, informs public policy, and is crucial to measuring the efficacy of public policy. Yet it is not a simple repository of facts on which we can draw. It is an ongoing process of evidence gathering, discovery, contestation, and criticism. I will argue that an understanding of the nature of science and the scientific process should be the central goal for scientific literacy, rather than a grasp of specific (often disciplinary) facts. With this understanding of science as a backdrop, the paper then turns to modes for citizen engagement with science. This paper articulates different ways citizens can engage with science, including four avenues for legitimate contestation of scientific claims. I then look more closely at contestation of science on the basis of values. That science can be legitimately contested by non-experts on a range of grounds means that science communication should not just aim at getting citizens to accept scientific claims, but rather to engage in a more robust two-way conversation about science.

6.1 Introduction[1]

Science is an important, often crucial, aspect of contemporary society. But its relationship to citizens is fraught. Even though scientists are among the most trusted professions generally (Smith and Son 2013), controversies about science in the

[1] My thanks go to the editors for their work on this volume (a splendid idea!) and for providing thoughtful comments on an earlier draft of this paper, to Ted Richards for reading and making this paper better, and most of all to Peter Machamer, for his incisive comments whenever we talk philosophy. Peter has an amazing ability to be both critical and supportive at the same time, which tends to bring the best out of people. His ability to draw from a wide range of sources, following inquiry wherever it leads, across whatever disciplinary boundaries, is part of his enduring stamp, and I hope it is reflected in this essay. My work on science and values began with his willingness to support an unorthodox dissertation project on science and policy, without which none of it would have happened.

H. Douglas (✉)
University of Waterloo, Waterloo, ON, Canada
e-mail: Hdouglas@uwaterloo.ca

M.P. Adams et al. (eds.), *Eppur si muove: Doing History and Philosophy of Science with Peter Machamer*, The Western Ontario Series in Philosophy of Science 81, DOI 10.1007/978-3-319-52768-0_6

public realm abound and scientists frequently lament the lack of scientific understanding the public exhibits.

Given this complexity, how should the public understand science and how should they relate to scientific endeavors? And, what kind of relationship with the public can scientists reasonably expect? In the realm of science communication since Sputnik, the predominant concern has been couched in terms of scientific literacy. Within the frame of scientific literacy, the central problem is that the public simply does not know enough about science, particularly scientific facts, and so is perpetually unable to grapple with scientific issues of public import. The "deficit" model (named for the deficit in scientific literacy extant in the public) is pervasive; it has also sustained substantial critiques in the past two decades. (Miller 2001; Bauer et al. 2007) Nevertheless, the prevalent attitude among scientists is that the public does not have a sufficient grasp of the underlying science to deal adequately with science-related policy issues (Pew Research Center 2015).

Critics, however, suggest that instead of deficits and one-way communication, there should be more two-way communication and more collaborative practices between scientists and citizens, to improve the conduct of science and public engagement with science. Such two-way communication can alert scientists both to the concerns of the public and to potentially erroneous assumptions in scientific work. It can allow for increased mutual understanding and respect among scientists and citizens. It can provide avenues for effective engagement in the practice of science and for deeper understanding of science among members of the public. But it can also open science up to scrutiny and critique by the public.

The possibility of a public willing to both engage with and critique science may be worrisome to many scientists. After all, such a description seems apt for perpetual thorns in the side of the scientific community, such as creationists and vaccine deniers (in particular, those who believe vaccines cause autism). Thus, we must have a way to assess whether a public's criticism of science is legitimate. What are the legitimate bases for being critical of science? And how can the public come to understand what the limits are on what is, and is not, a legitimate critique of science?

Here I provide an overview of the bases on which a citizen, in particular a non-scientist, can engage with and even criticize a scientific community or finding. While scientists might not want to open themselves to increased criticism, this is the price of genuine dialogue and engagement. If you want to talk to someone, you should expect that they might want to talk back. But there must also be a clear understanding of the basis on which a citizen can talk back to a scientist. What are the legitimate (and illegitimate) issues which can be raised? Delimiting these might help scientists see both what can be gained from engagement with the public and assuage concerns about what is on the table for contestation.

To see where the public can legitimately engage with and critique science, we need first a general account of what science is. This is not just useful for the particular philosophical project of finding the legitimate avenues of public contestation of scientific claims. It is also essential for the public to have this understanding of the nature of science and of the scientific enterprise, so that they can know what to reasonably expect from science and scientific experts. If we are to have two-way lines of communication, citizens need this understanding.

This paper begins with what we should think the nature of science is and what we should teach the public regarding science. The paper then presents various avenues for engaging with and, in some cases, critiquing scientific claims. Delving more deeply into those bases, the paper then turns to the role of values in science. Understanding the *legitimate* roles for values in science is important because if the public holds different values, particularly social and ethical values, from scientists, then where and when these values play a role in scientific practice shows important ways that the public can legitimately contest scientific claims.

6.2 Teaching the Nature of Science

For the past several decades, scientific literacy measures have aimed primarily at scientific facts. Although a few questions may address questions of scientific method, most are simple true/false questions that query whether citizens have a grasp of basic facts as science now sees them. Questions regarding the age of the planet, the size of electrons vs. atoms, whether all life has DNA in it, the speed of sound vs. light, and so forth, are the bulk of scientific literacy questions. Citizens have performed steadily poorly on these tests since their inception over 50 years ago, i.e. the public has been measured as being illiterate, and stably illiterate, over decades (Miller 2004).

Such results create despair among science communicators. The worry is that a scientifically illiterate public can neither appreciate science nor engage in policy debates with scientific components. But what if we are aiming at and measuring the wrong thing when attempting to assess scientific literacy? Our K-12 science courses still aim for a grasp of scientific facts—educational testing in multiple-choice format seems to be on the rise rather than the decline. The "answer in the back of the book" approach, while criticized at the national level in discussions of science education (e.g., AAAS's Project 2061, Schweingruber et al. 2012), does not appear to be on the wane at the level of implementation. Indeed, I have been told by educators that while national educational reform goals are great goals, getting them to be reflected in actual curricula is extremely difficult. The structural inertia seems insurmountable.

I want to argue here that actually achieving educational reform is crucial. And that educational reformers should be willing to jettison the teaching of large areas of facts in science for the sake of enabling citizens to understand *what science is*. If high school students take one course in science, it should not be chemistry or biology or physics, but rather a course in scientific investigation, so that they come to understand what science is like as an epistemic endeavor. It is this understanding they will need as citizens. All later understandings and facts can be built on this basis. If citizens don't have this basis, they will not be able to grapple with new scientific information as it arises throughout their lives.[2]

What should such a course teach? The most important thing to understand about science is its jointly critical and inductive nature. Science is an evidentially based endeavor that seeks to build an empirical understanding of the world. It does so through proposing explanatory theories that encompass the available evidence, and then, further tests those theories to the best of its ability. Because the theories always say more than the available evidence, the evidence provides at best inductive and thus incomplete support for the theories. Even simple claims of regularities have this feature—extending the claim of regularity into the past or future states, or new locations, requires an inductive inference. Similarly, theoretical or causal claims allow one to test such claims in new and different settings. These extensions then provide an opportunity to further test the theory. Thus, explanations and theories never have complete empirical support, yet the primary mode of support is empirical. It is in this sense that science is inductive.

That scientific claims are not completely supported by evidence allows for critical and ongoing testing of such claims. Because theories are applicable beyond their base of support, we can push on them, apply them to new contexts, to see if they continue to hold up. It is also this possibility for ongoing testing that makes scientific theories useful. If theories only applied to the already gathered evidence used to construct them, they would not be terribly helpful in making decisions about what to do in new contexts, including in practical policy-relevant contexts. Scientists can (and should) test their theories to see if they continue to predict phenomena accurately in new but relevant contexts. In addition, scientists can propose new theories that they think do better with respect to the available evidence. Because the fit between theory and evidence is never perfect and complete, space for ongoing critical activity is created. This is the work of scientists, to continually test, refine, reformulate, and rework theories and their relationship to evidence. Every claim of science is open to testing and reworking, even possible rejection. Scientific knowledge is, in this sense, all up for grabs.

As social epistemologists have noted, it is crucial that the social conditions of science be conducive for such critical activities. (Longino 1990, 2002; Solomon 2001) There needs to be open forums for critical exchange, open venues for airing ideas and results, expectations that criticisms are responded to, and an encouragement for diverse participants in the scientific community (so that new perspectives on old problems can be employed and so that tacit assumptions can be challenged). If the culture of science were one that discouraged criticism (e.g. that saw criticism as rude), it would not matter that in theory critiques of scientific claims could be offered. They would not be offered much in practice. The practice of science, of rewarding the critic, of encouraging new challenges and tests of old theories, is crucial to science's epistemic robustness.

---

[2] If this understanding of science can be taught effectively before high school, so much the better. Blackawton et al. 2011 suggests that it can.

Currently, the culture of science is not perfect in this regard. There are worries that there are too few incentives for replication studies (which try to show whether a theory holds up in a new instance), too much pressure for novel claims rather than building on and testing existing claims, too much pressure for winning grants from committees with an aversion to genuinely new ideas. Note that these concerns pull in opposite directions: that science is averse to the new and that science is obsessed with the new. The question in practice is where the balance is struck. More positively with respect to the robust social conditions for science, the scientific community has gotten a bit more diverse in the past few decades. (e.g., Shen 2013) More voices are now able to participate and raise scientific criticisms within the scientific community, even if more needs to be done in this regard.

It is crucial that citizens understand how the culture of science undergirds the practice of science. Raising concerns about the practicing culture of science is one legitimate way in which citizens can critique scientific claims. Citizens can also raise criticisms of scientific claims on the basis of the evidence, or alternative theories of the evidence, but doing so in practice is a challenge, as such criticisms must reflect an effective grasp of the currently available evidence, a grasp few non-experts have. As we will see below, there are other routes through which citizens can raise criticisms of scientific claims.

Courses should teach this understanding of the nature of science starting in grade school. Young students could engage in a community science project, grappling with data collection, controlling for confounders, proposing alternative explanations for phenomena, and seeing the competitive nature of science in practice in their classroom. (See, e.g., Blackawton et al. 2011) Alternatively, courses could track historical examples of scientific controversies and how they were handled and debated in the scientific community, including discussing the different theories that were proposed and debated, and how controversies were ultimately resolved. Students coming out of such courses would perhaps know fewer facts about a range of scientific topics, but have a deeper understanding of what it means to do science and the epistemic stance required to do it well. This is the understanding citizens should have first and foremost.

Why is understanding the nature of science more important than having a grasp of a range of scientific facts? Why have literacy tests been measuring the least important thing one can know about science? I am presuming that what we want from citizens is to be able to engage effectively with science policy disputes and to be able to assess scientific expertise. If this is the case, then teaching citizens scientific facts as the primary effort in school is not helpful.

Why? First, because the facts can change. When I was growing up, it was a fact that acquired characteristics could not be inherited by complex organisms. Because of new findings regarding epigenetics, this "fact" is no longer true. Science can challenge and change some of its core findings, even supposedly fixed facts. That it has this capacity is one of central reasons it is one of our most effective ways at discovering empirical reality. But that it has this capacity means that no scientific fact should be considered permanently stable. They are all potentially changeable by future evidence and discovery, even as they are also the best empirical account we have at any given time.

Second, if we teach citizens science as if it were a collection of stable facts to which we may add a few more, citizens will be disoriented and disappointed when either experts disagree or experts change their minds. That experts can disagree is to be expected in many cases—scientists disagreeing with each other is crucial to the culture of contestation in science. Experts changing their minds is also evidence of science *functioning properly*, not evidence of experts being fickle or weak-minded. Experts *should* change their minds in face of new findings, particularly if such findings cannot be accommodated by previous theories or are unexpected and replicated. Such changeability may be frustrating to citizens who just want to know what the right answer is, but that citizens have this desire and expectation just indicates how severely our science education system has failed them. Giving us the right answer, forever fixed and true, is not what the practice of science is about. Instead, it is about giving us the best, most empirically supported answer we can have at the time.

Third, when new science policy issues arise, such issues are often based on new scientific findings. Such new findings are not what is taught in school, and thus the schooling does little to serve as a "basis" for understanding the new findings. It might help a bit to have, for example, a clear sense of what a chemical compound is in order to understand what is at issue in a case of a local chemical spill, but more likely the details of the chemical spill, including the various plausible ecological pathways (e.g., uptake by which organisms), fat vs. water solubility, decomposition routes, and different remediation strategies were not part of the school chemistry, biology, or physics curriculum. And those issues not central to school curricula will be the crucial issues in the particular case. It makes more sense for people to be able to learn scientific facts as they need them for the particular case at hand. They will be more motivated to do so (as the relevance is clear) and which facts are important will be clearer as well.

For these three reasons, a curriculum structured towards teaching scientific facts, even scientific theories, for their own sake is inappropriate. The primary aim of science courses K-12 should be to teach what science is and how it works. Disciplinary content, specific scientific claims, facts, and theories, are all secondary to this goal. We should be structuring our science education accordingly.

6.3 Citizen Engagement with Science

Suppose we had such a citizenry educated in the nature of science. What would this enable? According to many polls, scientists tend to think that if citizens had a better grasp of science (usually meaning scientific facts), then citizens would agree more with scientists on central science policy issues like whether to pursue stem cell research or whether GMOs are safe. I disagree.[3] Not only do I think that a scientifically literate public (in the sense described above) would not necessarily agree with scientists more, but I also think that such a public could and should raise important challenges to scientists. Citizen engagement with science means more than the public coming to agree with scientists; it means the potential for citizens engaging critically with science, albeit on non-expert grounds.

The most widely understood sense of citizen engagement is through "citizen science" endeavors. (Dickinson and Bonney 2012) In these endeavors, citizens help scientists, usually by collecting data or by attempting to solve puzzles (like Foldit). (A full range of ongoing citizen science projects can be explored at sites like http://www.scientificamerican.com/citizen-science/.) In these cases, the agenda is generally set by the scientist and the citizen agrees to help out, whether by counting birds, keeping track of a temperature measure, or engaging in a constrained search.[4] Clearly, understanding the nature of science would help with these endeavors, but because the agenda (and often the methods) are set by scientists, citizens need only do what the scientists ask them to do, and thus an understanding of the nature of science is not actually necessary. Thousands of citizens participate in science through this avenue, but citizen science so conceived does not grapple well the fraught science policy contexts that drive the concerns over science literacy, nor does such an avenue invite a two-way discussion. The possibilities for citizen engagement with science do not end with citizen science.

Of deeper concern to both scientists and policy-makers are controversies regarding science (such as climate change, GMOs, and vaccines) where the scientists and citizens do not seem to agree about what to make of the available evidence. Scientists are perennially frustrated at the lack of concordance between citizen views and scientist views on these issues, and it is here that the deficit model gains the most traction as an explanation for why there remains a gap between scientist understanding and citizen understanding.

But the deficit model is untenable. Not only are deficits of understanding bi-directional, but in the case of controversial science, recent studies have found that increased scientific literacy (in the traditional fact-based sense) does not correlate with increased agreement with the views of scientists. In fact, in some cases, quite the opposite is true. For example, both Kahan et al. 2012 and Bolsen et al. 2015 have found that increased traditional scientific literacy does not correlate well with increased acceptance of expert consensus on climate change. For some portions of the public, the more literate the person is, the less they agree with scientists (or the more confident they are in their disagreement).

Much of the social science literature on differences between scientists and the public explain the phenomenon in terms of a "motivated reasoning" model. The

[3] Similar concerns are raised by Bauer et al. 2007, who note that there is empirical disagreement on whether "the more you know, the more you love it" or "familiarity breeds contempt." (p. 84)

[4] There are exceptions. See, e.g. Trautmann et al. 2012 for how citizen science can transform education contexts by allowing students to pursue inquiry with their own questions. Such a use of citizen science feeds well into the educational goals articulated above.

argument is that people are less interested in getting at the truth of the matter (to the best of their ability) and more interested in solving practical problems, such as deciding "which stances towards scientific information secure their personal interests." (Kahan et al. 2012, 733) But the kinds of reasons given under the umbrella of motivated reasoning (e.g. confirmation bias, peer group conformity, ideological commitments, and cherry-picking to fit with worldviews) are hardly the kind of reasons people would want to own up to publicly. It is in this sense that motivated reasoning looks less than rational.[5] Even if such cognitive tendencies explain the patterns of acceptance found, they do not justify them. And pointing to such explanations as reasons for why people diverge from experts is insulting to the members of the public who fit these patterns, precisely because the explanatory causes are not justificatory.

Thus, we currently have two predominant explanations for public disagreement with science, a deficit model and a motivated reasoning model (which explains why those who seem scientifically literate still disagree with experts), both of which tar the public with the taint of inadequacy. Both of these models presume that the goal is to get the public to agree with scientists, that whatever science communication system we have, it is working if we achieve this. But such a presumption raises the question of whether citizens can disagree with scientists for good reasons. What would good, sound disagreement with scientists look like?

Here I distinguish among four bases for disputing scientific claims. One can reject a scientific claim because (1) one rejects that the claim should be empirically based, i.e., one bases a claim on faith; (2) one disputes the empirical basis of the claim; (3) the scientists who developed the claim were not behaving properly, either individually or with respect to their epistemic community; and (4) one disputes the values that were a legitimate part of the development of the claim.

For the purposes of this paper, I will set aside disputations of claims based on faith. The epistemic stance of faith-based claims is qualitatively different from the epistemic stance of science-based claims. (Douglas 2015) Every scientific claim is open to challenge and rejection as new evidence is pursued, as noted above, whereas the point of many faith-based claims is to hold them above the fray of ongoing challenge and refutation. As faith-based claims are not open to evidential challenge (until, perhaps, the evidence is overwhelming), science has little to say to those who hold such claims. The evidence is largely irrelevant. Whether rejecting science on the basis of faith is rational is far too large of a topic to tackle here.[6] Minimally, what one should not do is conflate belief based on faith with belief based on evidence. This is precisely the error of those who want to teach creationism as science.

[5] A similar taint of irrationality is found in blaming the fragmented nature of the media and the way in which our information searches are structurally being driven to like-minded sources as a reason for why we disagree. Even if the information infrastructure distorts our searches in this way, we should act to counter such trends. (Miller and Record 2013)

[6] One classic instantiation of the debate is between Clifford and James. See Clifford (1877) and James (1896).

For the second basis, the evidence is central. There are cases where citizens have access to particular empirical information that experts do not, and thus disagreement with experts can be legitimately based on a purely empirical matter. (See, e.g. Wynne's classic Cambrian sheep farmer example, where experts had the wrong empirical understanding of grazing habits. Wynne 1996, 26) But such cases are rare, and often if we look at purely empirical bases for disagreement, citizens have anecdotal information whereas experts usually have a more complete empirical picture.[7] This is part of their basis of expertise. A more common source of disagreement concerns whether the right values are at play in the science—in whether the research agenda is pursuing the right questions and in whether what counts as sufficient evidence for the experts is also accepted as sufficient evidence by the public. More on this in the next section.

The third basis is often of central concern to the public. Allegations of fraud are supposed to be caught internally to the scientific community. That Andrew Wakefield's fraud in the case of autism and the MMR vaccine had to be caught by a journalist is something of an embarrassment. (Deer 2011) The scientific community is often not as effective at self-policing as it should be. In addition, the public can be worried that critics of the status quo are not being taken seriously enough within the scientific community. The strong resistance to the idea that bacteria can cause of ulcers is an exemplar of the occasional conservatism of the scientific community. (Thagard 1998) Sometimes, the seemingly outlandish idea is correct. Because of this possibility, the scientific community should address criticisms and alternatives, even if doing so is time consuming.

It is on the fourth basis that I will focus the remainder of the paper.

6.4 Social & Ethical Values in Science: Research Agendas and Inductive Risk

Values play a crucial role in scientific reasoning in at least two locations: (1) in the decision of which research projects to pursue (and the details of how to pursue them) and (2) in the assessment of whether the evidence is sufficient to support a claim. Both of these roles are legitimate for social & ethical values in science. Both of these roles also create locations for disagreement with, or contestation of, scientific work. I will illustrate each.

1. Research Agendas and Vaccines

Members of the public can dispute scientific claims because they think scientists are asking the wrong questions. As noted above, social and ethical values legitimately shape the attention of scientists to certain topics or questions. But if what

[7] Rare but often astounding and important, as can be seen in the recent case of Flint water contamination. See the story of Lee Anne Walters covered here http://michiganradio.org/post/mom-helped-uncover-what-was-really-going-flint-s-water#stream/0

scientists care about asking does not align with what members of the public care about knowing, statements based on the findings can be greeted with skepticism, because the public thinks the scientists are not answering the crucial questions.

One example of the mismatch between the public and scientists concerns vaccine safety. The Wakefield case and the purported link between vaccines and autism has created a great deal of harmful controversy. Wakefield's research was fraudulent, and there is no evidence for a link between vaccines and autism, despite careful searches for such evidence (Deer 2011, IOM 2004).

Nevertheless, even among parents who reject the idea that vaccines cause autism there is a hesitancy regarding vaccination schedules. (Goldenberg 2016) There are lingering concerns about side effects of particular vaccines or overwhelming the body with too many vaccines at once. Many public health officials are frustrated by this reluctance among parents to follow vaccination schedules.

One reason for such hesitancy involves the contemporary culture of parenting. As Maya Goldenberg has noted (Goldenberg 2016), parents are told that in general, they should closely monitor their children, that different children have different needs, and that if they are paying attention to their children, they will know their children best. In the face of this individualistic parental expertise culture, parents are asked to set aside their individualized knowledge and follow the herd in the case of vaccination schedules. We should not be surprised that many parents balk at this.

More centrally for our purposes here, many of the questions parents have about vaccines cannot be answered well because they have not been extensively researched. While general safety and efficacy levels of vaccines are studied, what causes the rare serious side effect (which do exist) is not well understood. And many side effects are not tracked well at all. If a child has a response once at home, it is rarely recorded or taken seriously. Such anecdotal evidence is not carefully collected or examined, making the search for harmful side effects of vaccines less than robust. (Stegenga 2016) When parents ask what makes a child susceptible to such adverse effects, physicians have no good answers. It is simply not part of their research agenda. The emphasis in vaccine research has been in producing a generally safe and effective product to produce herd immunity. What causes adverse effects in the rare case is not well understood or studied.

Thus, if parents do see an adverse event in their child at home, they will not receive useful information from their doctors about what it means for the vaccination schedule in general. Nobody knows. It is not surprising that hesitancy to trust the schedule is born out of this mismatch of concern: the physicians with population level trends and the parent with individual response. As a result of this mismatch, parents have good reason to be hesitant placing full trust in the experts—the experts do not have the expertise parents desire.

Similar concerns can be raised regarding the environmental safety of some GMOs and regional climate forecasts. In such cases, what some citizens think are the crucial questions have not been well studied (as of yet). The values of those citizens and the values of the scientists are not aligning, producing skepticism about what scientists are reporting. The scientists are not answering the questions on the topic that are of import to the skeptical citizens.

6.5 Neonicotinoids and Inductive Risk

In addition to values shaping the research agenda, there is the issue of whether the evidence we have is sufficient for our acceptance of (and action on) the claim. This is a pervasive yet subtle role for values in science. It is also a crucial reason for why it can be perfectly rational for members of the public to trust experts who share their values.

Inductive risk arises whenever knowledge is inductively based (there is thus a chance of getting it wrong–whatever the knowledge claim is) and there are clear consequences to getting it wrong. In short, inductive risk is endemic in science. In societally relevant areas of science, inductive risk opens the door to social and ethical values in the assessment of what counts as sufficient evidence for a claim. Depending on which consequences of error you find more acceptable, you can be legitimately concerned about some kinds of error (e.g., false positives) more than other types of error (e.g., false negatives). The assessment of the consequences of error is where the values play a role (what I have called an *indirect role*). (Douglas 2009) One must assess such consequences because there is no fixed threshold for what counts as sufficient evidence (e.g., what counts as statistically significant varies among fields).

Now, when considering values and inductive risk, the values do not dictate a result or even provide a reason for choosing a particular result. What they do is provide a reason for setting evidential standards in a particular place. They help to decide whether the evidence is enough. In W.K. Clifford's terms, they help set what should count as "sufficient evidence." (Clifford 1877) Thus, evidence maintains a central place in the reasoning structure (unlike with faith-based belief). Evidence must also be present and substantive (values cannot make up for a clear lack of evidence). But rational people can still disagree about whether the evidence available is strong enough or good enough.

Live science policy debates illustrate inductive risk best. Consider the debate over whether bees are threatened by the use of neonicotinoids. There are some studies which suggest they are. The class of pesticides is immunological and neurotoxic to all insects, and problems with bee colonies correlate with the use of the pesticides (although it takes a few years to show up). But is the dose bees experience in the environment enough to cause them harm? Some have argued that at current levels of use, the dose is not high enough, that studies showing harm have used excessively high doses, and so farmers should be allowed to continue using the pesticides as they have been. (e.g. DEFRA UK 2013) Others argue that controlled studies have been replicating field realistic doses and that critics of the studies have not been taking into account exposure routes like guttation (EFSA EU 2013).

In short, there is expert debate. What is the public to do? It makes sense in these cases to listen to experts that share your values. This is because experts that share your values will assess the consequences of error and the sufficiency of evidence as you would. If you are particularly worried about bee health, then scientists who are similarly worried about bee health (and thus perhaps willing to see less evidence as being sufficient before taking regulatory action) would be most trustworthy for you. (Wilholt 2013) If you are particularly worried about farmers minimizing pest damage, you should trust scientists with similar values, for they will demand more evidence before accepting the claim that neonicotinoids harm bees.

What science communicators have noted,[8] that sharing core values with an expert increases the uptake of what the expert says, is not just a social bias, an irrationality that we can potentially exploit once we know about it. It is a rational strategy[9] when faced with inductive risks. As inductive risks are pervasive in science, it is a rational strategy when dealing with scientific claims.

In addition, pointing out the values that undergird different assessments of evidential sufficiency help foster good debate within science rather than undermine scientific practice. That such divergent values and divergent assessments can exist should help focus scientists where their efforts should be focused—on seeing if they can produce better, stronger evidence to convince skeptics (both among their colleagues and among the public). And public debates can consider both the available evidence and the arguments for why some values should be more important than others.

6.6 Conclusion

There are multiple ways in which citizens can legitimately contest scientific claims, even within an empiricist framework (i.e. setting aside faith-based claims). They can do so on an empirical basis, though gathering evidence for the non-scientist is challenging. They can critique the way in which criticisms are addressed and the interchange of ideas within the scientific community. Finally, they can query whether a research project is well-conceived, whether it has an acceptable (either morally or epistemically) methodology, and whether the evidence gathered is sufficient for the claims being made by scientists. Citizens can rationally and legitimately prefer to depend on the expertise of those scientists who share relevant moral and ethical values.

This means that science will be a site of debate and critique as long as its implications are relevant to the public and to policy-making. Science is not a closer, not a mechanism for bringing about acquiescence or unanimity among the public. Science instead is a resource for debate and a resource for imagining our futures in an already messy public realm.

That does not mean science can and should be used to support any position whatsoever. Scientific evidence places limits on what is empirically plausible, and if we are to care about evidence, those limits must be respected. (Douglas 2014) Scientific integrity still matters, and that the political realm can and does use science as a

[8] As Liz Neeley notes Achenbach 2015, 45.

[9] By rational, I don't just mean in one's practical interests. I also mean justifiable publicly, i.e. a reasoned basis that can be stated publicly.

resource in debates does not mean it should be a resource bent, infinitely flexibly, to the will of the user.

But science is also not an inflexible producer of permanent truth. The conception of science as a fount of fixed facts is as problematic a conception of science as infinitely flexible, as a source of whatever view you desire. Citizens need to understand the ongoing nature of science, the way in which evidence constrains it, but also the way in which different interpretations of the evidence are also likely. With this understanding, productive and respectful engagement between scientists and the public becomes possible.

References

Achenbach, J. 2015. The Age of Disbelief. *National Geographic,* March Issue, 227: 30–47.

Bauer, M.W., N. Allum, and S. Miller. 2007. What Can We Learn From 25 Years of PUS Survey Research? Liberating and Expanding the Agenda. *Public Understanding of Science* 16(1): 79–95.

Blackawton, P.S., S. Airzee, A. Allen, et al. 2011. Blackawton Bees. *Biology Letters* 7(2): 168–172.

Bolsen, T., J.N. Druckman, and F.L. Cook. 2015. Citizens', Scientists', and Policy Advisors' Beliefs About Global Warming. *The Annals of the American Academy of Political and Social Science* 658(1): 271–295.

Clifford, W.K. 1877. The Ethics of Belief. In *The Ethics of Belief and Other Essays*, ed. T. Madigan, 70–96. Amherst: Prometheus.

Deer, B. 2011. Secrets of the MMR Scare: How the Case Against the MMR Vaccine Was Fixed. *British Medical Journal* 342: c5347.

Department for Environment, Food, and Rural Affairs (DEFRA). 2013. *An Assessment of Key Evidence About Neonicotinoids and Bees*. https://www.gov.uk/government/publications/an-assessment-of-key-evidence-about-neonicotinoids-and-bees.

Dickinson, J.L., and R. Bonney. 2012. *Citizen Science: Public Participation in Environmental Research*. Ithaca: Cornell University Press.

Douglas, H. 2009. *Science, Policy, and the Value-Free Ideal*. Pittsburgh: University of Pittsburgh Press.

———. 2014. Scientific Integrity in a Politicized World. In *Logic, Methodology, and Philosophy of Science: Proceedings of the Fourteenth International Congress*, ed. Peter Schroeder-Heister, Gerhard Heinzmann, Wilfrid Hodges, and Pierre Edouard Bour, 253–268. London: College Publications.

———. 2015. Politics & Science: Untangling Values, Ideologies, and Reasons. *The Annals of the American Academy of Political and Social Science* 658: 296–306.

European Food Safety Authority (EFSA). 2013. *EFSA Guidance Document on the Risk Assessment of Plant Protection Products on Bees* (*Apis mellifera*, Bombus spp. and solitary bees). http://www.efsa.europa.eu/en/efsajournal/pub/3295.

Goldenberg, M. 2016. Public Misunderstanding of Science? Reframing the Problem of Vaccine Hesitancy. *Perspectives on Science* 24: 552–581.

IOM (Institute of Medicine– Immunization Safety Review Committee). 2004. *Immunization Safety Review: Vaccines and Autism*. Washington, D.C.: National Academies Press.

James, W. 1896. *The Will to Believe: And Other Essays in Popular Philosophy*. New York: Longmans, Green & Company.

Kahan, D.M., E. Peters, M. Wittlin, P. Slovic, L.L. Ouellette, D. Braman, and G. Mandel. 2012. The Polarizing Impact of Science Literacy and Numeracy on Perceived Climate Change Risks. *Nature Climate Change* 2(10): 732–735.

Longino, H.E. 1990. *Science as Social Knowledge: Values and Objectivity in Scientific Inquiry*. Princeton: Princeton University Press.

———. 2002. *The Fate of Knowledge*. Princeton: Princeton University Press.

Miller, S. 2001. Public Understanding of Science at the Crossroads. *Public Understanding of Science* 10(1): 115–120.

Miller, J.D. 2004. Public Understanding of, and Attitudes Toward, Scientific Research: What We Know and What We Need to Know. *Public Understanding of Science* 13(3): 273–294.

Miller, B., and I. Record. 2013. Justified Belief in a Digital Age: On the Epistemic Implications of Secret Internet Technologies. *Episteme* 10(02): 117–134.

Pew Research Center. 2015. *Public and Scientists' Views on Science and Society*, January 29, 2015.

Schweingruber, H., T. Keller, and H. Quinn, ed. 2012. *A Framework for K-12 Science Education: Practices, Crosscutting Concepts, and Core Ideas*. Washington, D.C: National Academies Press.

Shen, Helen. 2013. Inequality Quantified: Mind the Gender Gap. *Nature* 495**:** 22–24. http://www.nature.com/news/inequality-quantified-mind-the-gender-gap-1.12550.

Smith, T., and J. Son. 2013. *General Social Survey 2012 Final Report: Trends in Public Attitudes About Confidence in Institutions*. Chicago: NORC at the University of Chicago.

Solomon, M. 2001. *Social Empiricism*. Cambridge, MA: MIT press.

Stegenga, J. 2016. Hollow Hunt for Harms. *Perspectives on Science* 24: 481–504.

Thagard, P. 1998. Ulcers and Bacteria I: Discovery and Acceptance. *Studies in History and Philosophy of Science Part C: Studies in History and Philosophy of Biological and Biomedical Sciences* 29(1): 107–136.

Trautmann, N.M., J.L. Shirk, J. Fee, and M.E. Krasny. 2012. Who Poses the Question? Using Citizen Science to Help K-12 Teachers Meet the Mandate for Inquiry. In *Citizen Science: Public Participation in Environmental Research*, ed. Dickinson and Bonney, 179–190. Ithaca: Comstock Publishing Associates.

Wilholt, T. 2013. Epistemic Trust in Science. *The British Journal for the Philosophy of Science* 64(2): 233–253.

Wynne, B. 1996. Misunderstood Misunderstandings: Social Identities and the Public Uptake of Science. In *Misunderstanding Science? The Public Reconstruction of Science and Technology*, ed. A. Irwin and B. Wynne. Cambridge, MA: Cambridge University Press.

Chapter 7
Political Philosophy of Science in Nineteenth-Century France: From Comte's Positivism to Renouvier's Conventionalism

Warren Schmaus

Abstract Recent controversy over whether the Vienna Circle can provide a model for today's political turn in the philosophy of science indicates the need to clarify just what is meant by the term political philosophy of science. This paper finds fourteen different meanings of the term, including both descriptive and normative usages, having to do with the roles of political values in the sciences, the political consequences and significance of the sciences and scientific modes of thought, and political processes within the sciences. These different senses are exemplified by the history of nineteenth century French philosophy of science, including Comte and Renouvier, as well as twentieth century logical positivists such as Neurath and more recent philosophers of science such as Kitcher and Longino. The paper also addresses the issue of how there could be a positivist political philosophy of science, given positivisms' questioning of values, and considers Renouvier's social contract approach to a political philosophy of science more promising than either Comte's naturalism or the logical positivists' non-cognitivism.

7.1 Introduction

Political philosophy of science is a relatively new term. It was perhaps first used a generation ago by Joseph Rouse (1987), but began to be discussed critically in the literature more frequently about a decade ago (Gomez 2004; Uebel 2005). Philosophers often invoke the term *social* when discussing the wider context in which science is practiced. But *political* philosophy of science suggests something different: a concern with power relationships, the structure of government, and party politics. So what could a political philosophy of science be? There does not seem to be a single answer.

W. Schmaus (✉)
Department of Humanities, Lewis College of Human Sciences, Illinois Institute of Technology, Chicago, IL 60616, USA
e-mail: schmaus@iit.edu

M.P. Adams et al. (eds.), *Eppur si muove: Doing History and Philosophy of Science with Peter Machamer*, The Western Ontario Series in Philosophy of Science 81, DOI 10.1007/978-3-319-52768-0_7

One of the earlier uses of the term is by Thomas Uebel (2005), who argues that political motivations were present at an earlier stage in the philosophy of science, in particular among the members of what has come to be called the Left Vienna Circle (LVC): Otto Neurath, Rudolf Carnap, Hans Hahn, and Philip Frank. Sarah Richardson (2009a, b), on the other hand, has challenged the claim that the LVC's was a political philosophy of science and affirmed that they do not provide a model for a political philosophy of science today.

The presence of controversy is often a symptom that multiple meanings of a key term are afoot. I do not intend to settle the argument over what should count as a genuine political philosophy of science, but rather to analyze the different possible meanings this term could have. Getting clear about these possible meanings can help us avoid unnecessary disputes. We can allow for different sorts of intellectual pursuits under the general rubric of political philosophy of science. Such an analysis can also help us to recognize that although the term "political philosophy of science" may be new, the ideas collected under this term have a history. Some of its possible meanings can be brought out in sharper relief through comparisons among not only the Vienna Circle logical empiricists and more recent philosophers of science, but also such nineteenth-century French philosophers as Auguste Comte (1798–1857) and Charles Renouvier (1815–1903).

Nineteenth century France may be particularly instructive because political motivations were behind the rise of the positivist philosophy of science there. Claude Henri de Rouvroy, Comte de Saint-Simon (1760–1825) invented the term positive philosophy as the name for the idea of creating a better society through the application of scientific ways of thinking. He conceived this philosophy as a response to the political and social turmoil that occurred in the wake of the French Revolution and Napoleonic wars. But Saint-Simon knew little about science, and it was Comte who developed a philosophy of science that he had hoped would advance the scientific re-organization of society. Renouvier was initially attracted to Saint-Simonianism and Comtean positivism but became increasing dissatisfied with such movements. His critical response combined arguments from political philosophy with arguments from the philosophy of science.

The link between political philosophy and positive philosophy of science would appear to be problematic, as positivism calls into question the meaningfulness of statements about values. A look at the past reveals that there have been different approaches to this problem, including not only the Vienna Circle's non-cognitivism but Comte's naturalism and Renouvier's non-naturalist alternative. I do not intend to put forth any of these philosophies as a model for a political philosophy of science today; I prefer to remain a pluralist about such matters. Nevertheless, these comparisons may help us to see what works and what doesn't work.

7.2 What Political Philosophy of Science Could Mean

Let's begin with the current controversy. According to Uebel (2005, 760), the LVC philosophers recognized that political values may play a role within science due to the underdetermination of theories by the evidence for them. On the other hand, Richardson (2009a, 23) has argued that, although the LVC may have been politically engaged, their ethical non-cognitivism would have prevented them from developing any sort of a political philosophy. In his reply, Uebel (2010, 217) stressed that the LVC did not *endorse* the appeal to political values in theory evaluation, and thus theirs was a political philosophy of science only in a descriptive and not a normative sense.

So far, then, a political philosophy of science appears to be one that allows some role for political values in theory choice, and this can be understood in either a descriptive or normative sense. However, there appear to be at least two more possible senses of the term political philosophy of science implicit in Uebel's original article, in which he also discussed what he calls the "superstructural role" of philosophy for the LVC:

> offering methodological reflections and developing logico-linguistic frameworks and procedural or institutional frameworks for possible use. Consistent with their conception of philosophy, the assistance that the scientific world-conception could give to practical ends could be but methodological: it could provide tools for clear thinking. (Uebel 2005, 758.)

That is, although the logical positivists may not have endorsed a role for political values in theory choice, there was nevertheless a connection between their philosophy and their politics. What linked the LVC's scientific world-conception to left politics was the idea of organizing society on rational principles, which depended on the tools for clear thinking that philosophers could provide (Uebel 2005, 758–760). The two different kinds of frameworks in the passage quoted from Uebel suggest two additional senses of political philosophy of science: One is a philosophy of science concerned with linguistic choices that could advance a political ideology through providing tools for clear thinking while the other is concerned with political organization.

There now appear to be three different but related categories that the various senses of political philosophy of science fall under. First, it can mean a philosophy of science that concerns the role of political values in scientific decision making. Let's call this **V** for values. Then there's a philosophy of science that considers the political implications or consequences of theories in science or scientific ways of thinking. We'll call this **C** for consequences. Finally, there is a philosophy of science that is concerned with political processes, institutions, and organization within the scientific community and in its relations with society at large. This one we'll label **P** for political process. Adopting the mathematical convention for the use of parentheses, **P** could be called a ((political philosophy) of science), whereas **V** and **C** might be better characterized as (political (philosophy of science)). In other words, **P** is like political philosophy applied to science, while **V** and **C** start from philosophy of science and then turn to political considerations. There are multiple and overlapping meanings within these three categories and each of them can be used in either a descriptive or a normative mode, although the normative typically presupposes the descriptive. In practice, **V** and **C** can occur together, when decisions are made because certain consequences are valued. However, **C**, at least in the descriptive mode, may also concern itself with the unintended consequences of science and its modes of thought.

Political philosophy of science in sense **V** can take on different meanings in accordance with the kind of scientific decisions that political values are said to influence. Using a small **d** and a small **n** to distinguish the descriptive from the normative use of the concept, we shall use the labels **Vd1** and **Vn1** for the two positions Uebel distinguished: a philosophy of science that simply recognizes the role of political values in theory choice and one that actually endorses this role. Both of these senses presuppose the underdetermination argument. Some philosophers of science have cast doubts on this argument (e.g. Kitcher 2001, ch. 3). But although I am sympathetic to these criticisms, my purpose here is not to evaluate the LVC's position, but only to draw some needed distinctions.

In their controversy, Uebel and Richardson also discussed a second kind of decision that may be influenced by political values, regarding what Uebel calls the development of a logico-linguistic framework. In the Vienna Circle protocol sentence debate, the question arose as to the role of political values in whether to opt for a physicalist or a phenomenalist language. Carnap appears to have been indifferent, regarding the choice as purely conventional. Neurath, on the other hand, subscribed to the normative philosophy **Vn2** that values *should* play a role. He thus preferred the physicalist language insofar as it would promote the sort of clear thinking that would advance the scientific world-conception, which would have political consequences (Uebel 2005, 759). **Vd2** would be the philosophy that just recognized that political values may guide scientists and philosophers in their choice of language. As Neurath was concerned with the political consequences of the scientific world-conception, his views also illustrate **Cd1** and **Cn1**, as I'll explain below.

Stepping aside from controversies over the LVC, we can distinguish another possible sense, **Vd3**, in which political values may influence choices of methodology. **Vd3** is implicit in Steven Shapin and Simon Schaffer's *Leviathan and the Air-Pump* (1985), in which they argued that Hobbes and Boyle differed over the role of experiment in science for political reasons. **Vn3** is at least implicit in feminist writings that advocate that medical research should not limit itself to the use of male experimental organisms, whether rodent or human. Also, Janet Kourany (2010, 6–7) finds that feminist philosophers of science tend to endorse epistemic values such as theoretical novelty, ontological heterogeneity, and complexity of relationship over more traditional values such as consistency with established knowledge, simplicity, and explanatory scope, arguing that the non-traditional values are more in line with egalitarian political goals. And epistemic values can influence the choice and direction of a methodology.

Political values may also shape the sciences by influencing research agendas. Both descriptive, **Vd4**, and normative, **Vn4,** senses of this understanding of a political philosophy of science are implicit in the works of some contemporary philosophers of science. Actual cases of **Vd4** and **Vn4** discussed in the literature involve different senses of **P** as well. For instance, Philip Kitcher in *Science, Truth, and Democracy* discusses how non-epistemic, political values and political processes have shaped the United States research agenda, favoring the Human Genomes Project but not the Superconducting Supercollider (2001, ch. 1). **Vn4**, which says that political values *should* play a role in setting research agendas, is at least implicit in Kitcher's arguments for an enlightened democracy to set a society's research priorities (2001, ch. 10).

Turning now from how values may shape science to the political consequences or significance of science and scientific ways of thinking, we label the simple statement that there are such consequences **Cd1**. The term political philosophy of science was used in the sense by Rouse (e.g. 1987, 252) and more recently in an article by Ambrosio Velasco Gomez titled "Toward a Political Philosophy of Science" (2004). In this article, Gomez discusses the ideas behind all three senses of political philosophy of science, **V**, **C**, and **P**, but without drawing further distinctions. He discovers what I have been calling **Cd1** in works by Paul Feyerabend, Hans-Georg Gadamer, and Jürgen Habermas, who have written about the consequences of what Neurath called "pseudo-rationalism," the philosophy that scientific knowledge is justified through purely rational, methodological procedures, that is, the denial of both **Vd1** and **Vn1**. These three philosophers Gomez cites have argued that pseudo-rationalism, together with the assumption that scientific knowledge can justify political decisions and legitimize a political order, has anti-democratic, authoritarian consequences. Together, these assumptions do not make room for the free participation of citizens in political life, since they lack the scientific expertise necessary for making policy choices. Of course, this undesirable consequence does not follow without the critical second premise. But as we shall see it is a consequence of Comte's political philosophy of science. **Cd2** presupposes but goes beyond **Cd1**, to unmask the extra-theoretical function that scientific theories and ways of thinking have served, since these functions will be found among the consequences of some theory or mode of thought. An example would be Feyerabend's argument that science has advanced white male interests at the expense of colonized peoples (1987, Introduction).

Unmasking can take place in a purely descriptive mode; it becomes normative only when some judgment, typically critical, is passed on these extra-theoretical functions. Then it is part of **Cn1**. But a **Cn1** could also endorse certain political consequences of science and scientific ways of thinking. The LVC provides an example, insofar as they promoted scientific tools for clear thinking that they believed would advance their political agenda, as we have seen. Popper's philosophy also illustrates **Cn1**, but in a different way. That is, Popper was concerned not with political consequences in any causal sense so much as with the *logical* consequences of his philosophy of science for political theory. Perceiving the fallible character of scientific knowledge, Popper endorsed epistemic values such as boldness in speculation and the willingness to criticize accepted theory. He then saw that these values in turn call for civil liberties such as freedom of thought and discussion

(Popper 1965, 352). **Cn2**, on the other hand, draws on philosophy of science in pursuing the methodological or foundational task of establishing or defending an alternative system of thought that can serve as the basis of a new political order. This sort of political philosophy of science is connected with scientific socialism, Saint-Simonianism, Comtian positivism, and other nineteenth-century ideologies. To twentieth-century philosophers, especially those Gomez cites, this looks like the road to an authoritarian political order.

The difference between **Cn2** and **Pn,** a philosophy of science concerned with political processes, is the difference between on the one hand drawing consequences from a philosophy of science for establishing a governing ideology for society (**Cn2**) and on the other hand theorizing about processes by which scientists and the general public may achieve a consensus about theories, methodologies, and research programs (**Pn**). As we shall see, it's the difference between the philosophies of Comte (**Cn2**) and Renouvier (**Pn**). When Kitcher provides an account of how we arrived at an elitist system of making science policy, he is engaged in **Pd** (2001, ch. 11), but when he puts forth his ideal of well-ordered science, he's engaging in **Pn** (2001, ch. 10). So is Helen Longino in *The Fate of Knowledge* (2002) and *Science as Social Knowledge* (1990), where she stipulates four norms that distinguish a genuine knowledge-producing community: public venues for criticism, public standards for criticism, uptake of criticism, and "tempered equality," that is, that everyone regardless of economic and social status should be able to express their point of view as a knowledge community works towards a consensus about what to accept as knowledge (2002, 129–133).

7.3 Comte's Political Philosophy of Science

Comte's positive philosophy was unlike that of the logical positivists in more ways than one. First, Comte took a naturalist rather than a non-cognitivist approach to questions of values. Thus he was able to forge a stronger link between his philosophy of science and his political philosophy. His was a political philosophy of science in several senses of both **C** and **P** but not in any of the senses of **V.** It exemplified **Cn1** by endorsing scientific ways of thinking and **Cn2** by applying them to the development of an organizing political ideology. For Comte, the positive philosophy was the final stage in the development of our knowledge, which would usher in the final political and social system. Engaging in a bit of sociological unmasking *avant la lettre*, Comte argued that the older, theological philosophy supported the Church, monarchy, aristocracy, and the military; while the metaphysical philosophy propped up the newer class of lawyer-politicians and industrialists. Thus Comte's philosophy could also exemplify **Cd2.** The positive philosophy could also be characterized as a **Pn**, albeit not an attractive one, as we shall see.

Comte's motivations for creating a philosophy of science were political and social. Like some recent philosophers, Comte was concerned with organizing science to serve human needs. But for him, this included not only practical technological applications of science but also replacing the Christian religion with a new intellectual consensus grounded in the sciences that would unify society and ensure order and progress. Social harmony, he thought, depended upon intellectual harmony: "the world is governed and overturned by ideas" (Comte 1830–1842, I, 38; 1988, 28). By the term positive philosophy, Comte meant a philosophy that would play a positive role in building a new society, in contrast with what he took to be the negative philosophy of the Enlightenment, which could only criticize and undermine the older regime of Church and monarchy but offer nothing stable to replace it. Drawing on the empiricist tradition, it also sought positive or certain knowledge of observable relations among phenomena, rather than unverifiable hypotheses about their hidden causes. These observable relations could serve practical needs, while theories of hidden causes served no useful purpose (1844, 41–43; 1851–1854 I, 58; IV, 547; 1875–1877 I, 45; IV, 473).

Although Comte was attracted by Saint-Simon's idea of society organized on scientific principles, he thought he had a better idea of what that entailed, having been educated in mathematics and the physical sciences at the *École Polytechnique.* In the six-volume *Cours de philosophie positive*, written from 1830 to 1842, he explained how the positive method developed first in mathematics and then in succession in astronomy, physics, chemistry, and biology, and was now ready to achieve its final stage in his new science of sociology, which was to provide the knowledge necessary for social and political reorganization. The positive method now made possible a knowledge of the natural laws governing social phenomena, allowing for the prediction ("prevision") of the future course of events, which could help rulers make policy decisions – something the theological and metaphysical methods did not make possible, as they attributed phenomena to the actions of arbitrary wills (1830–1842 II, 108). Subsequently, Comte added morality as a seventh positive science and worked out the particulars of his ideal society in the four-volume *Système de politique positive* (1851–1854).

Comte also appreciated the social character of scientific inquiry. He recognized that scientific knowledge is the collaborative product of a community of researchers, not the work of an isolated Cartesian genius as it was for his French contemporaries in academic philosophy. Scientists depended upon one another for critical discussion, and upon social conditions in the wider society that would allow their pursuits (Bourdeau 2006, 125–136). Yet in Comte's positivist utopia, scientists would debate merits of theories in accordance with positive methods, of which he believed he had given the final and definitive account in the *Cours.* Also, although he recognized scientists' need for liberty to debate each other's ideas, his political system provided no guarantees for it. Indeed, his philosophy undermined civil liberties. For Comte, the notion of free will belonged to the metaphysical stage of thought; people were determined in their actions by natural laws, mooting questions of political liberties.

Perhaps what is most valuable in Comte's political philosophy is the idea that our leaders need to be educated in the history, philosophy, and sociology of science, given the increasing role of science and technology in contemporary society. Less attractive are the details of how he would put this idea into practice. Comte reintroduced into political theory the distinction between temporal and spiritual powers – although he was aware of the misleading connotations of these terms (1851–1854 I, 214; II 314–315; 1875–1877 I, 173; II, 257–258). Unlike the former spiritual power, whose authority rested on revealed truth, the positive spiritual power depended on articles of the positive faith that were open to examination and experimental testing (1830–1842 II, 667–668). In the positivist system, the temporal powers were to be held by the captains of industry, but they were to be subordinate to the spiritual powers, to be held by a secular priesthood educated in the sciences, including his new science of sociology. These savants would see to it that society served the general interest, not just the interests of the temporal elite.

This new positivist spiritual power would take over control of education from the Church and introduce scientific ways of thinking to the masses. Recognizing that government should rest on public opinion and not on force, Comte thought all citizens, men and women, should receive an education in science and its methods. Thus for 17 years, Comte provided free public lectures at his own expense for the working classes on the history of astronomy, as a way of introducing them to scientific methods. Gomez (2004, 120) sees Comte as having tried to break the link between pseudo-rationalism and authoritarian politics by promoting scientific education in the general population. However, in his political philosophy, Comte provided no means by which this enlightened general public could hold their leaders accountable and exercise control over them (Bourdeau 2006, 129–130). Thus, although he may have intended the positivist political system to be progressive, he proposed no safeguards against it slipping into an authoritarian system in which the elite arrogate all power to themselves.

Ultimately, Comte's political message was unclear, inspiring movements on both the left and the right. The left was attracted to his replacing the traditional Church with a secular, scientifically educated intelligentsia. However, Comte dismissed familiar notions of political philosophy, such as individual rights and liberties, as belonging to metaphysical stage. For similar reasons, he rejected the idea of a social contract (Pickering 1993, 225). According to Comte, society was an organism that developed according to its own set of laws that sociology must observe and study, not invent. This notion spoke to the right's prejudice that social planning can never work.

7.4 Renouvier's Two-Pronged Attack on Comtean Positivism

Unlike Comte, Renouvier was much concerned with the defense of individual liberties. Renouvier's political philosophy of science differed from Comte's in two important ways. First, Renouvier did not attempt to construct a governing political ideology from his philosophy of science. Although his philosophy combined **Pn** with **Cd1**, **Cd2**, and **Cn1**, it did not include **Cn2**. Second, Renouvier recognized the limitations of Comte's naturalism. He saw that philosophy has normative content, which had to be defended with philosophical arguments. Comte had tried to replace philosophy with empirical science, which Renouvier thought was not up to the task. For Renouvier, morality was not and could not be a positive science. In fact, he realized that the sciences themselves depend on philosophical assumptions.

Like Comte, Renouvier was a *polytechnicien* and was self-taught in philosophy. He had already read the first volume of the *Cours de philosophie positive* before he began his studies at the *École Polytechnique* from 1834 to 1836, where Comte taught him mathematics. Renouvier became associated a group of socialists who had broken away from the Saint-Simonians, which after the death of Saint-Simon in 1825 had changed into an authoritarian religious cult under the leadership of Prosper Enfantin. As a young man, Renouvier was politically active with this breakaway group, which included Pierre Leroux, Hippolyte Carnot and Jean Reynaud, publishing political articles with them and participating in the provisional government set up in the February 1848 revolution (Fedi and Huard 2002, 25–26; Tresch 2010, 327). However, he became disgusted with practical politics after Napoleon III's coup d'état in December 1851, and retreated to his family estate in the south of France, where he began work on his major philosophical works, the *Essais de critique générale*, in which he articulated and defended Kantian metaphysical and epistemological positions.

After the *débâcle* of 1870–1871, France experienced a great variety of competing political ideologies, ranging from monarchism and imperialism through countless forms of Republicanism to socialism and communism. In order to help the young republic finds its direction, Renouvier and his collaborator François Pillon (1830–1914) founded the weekly *La Critique philosophique, politique, scientifique, littéraire* in 1872. This weekly addressed political issues in the new Third Republic from a philosophical point of view, including the defense of political and civil liberties against ideologies that sought to limit individual liberties in the interest of social stability and order. Renouvier was especially concerned with unmasking the pretension to scientific status of Saint-Simonianism, Comtian positivism, communism, and other social-political movements that claimed the status of science to give themselves the aura of truth (Renouvier 1873a). Thus his was a political philosophy in the sense of both **Cd2** and **Cn1**. For Renouvier, a political program that claims to be grounded in a completed science risks sliding into totalitarianism and tyranny. To unmask these ideologies entailed a philosophical investigation of just what it meant for something to be a science. Renouvier challenged anyone who would give the name "science" to some proposition that they hold true to tell us to what science it belonged, and what right it had to distinguish itself from hypothesis, "because everyone knows that each science applies to its objects of experience hypotheses that serve it as ways of working and furnish some sort of stations and resting places for the mind." He thought that those who abused the term "science" were incapable of meeting this challenge. Indeed, Renouvier questioned whether there even is such a thing as "science" and suggested that there are only *sciences*, each of which rests on a hypothesis that may be overthrown someday (1876, 401–402).

To defeat the positivist program, Renouvier undertook a critical examination of Comte's philosophy of science as well as his political philosophy. Even as early as his first *Essai* of 1854, in which he had defended a generally positivist interpretation of the sciences, he had criticized Comte's social philosophy as dogmatic and illiberal for its denial of individual rights (1912a, I, xvi). By the 1870s, he also found Comtean positivism wanting as both a descriptive and a normative account of science. Renouvier questioned whether Comte's *Cours* was au courant even in the sciences of Comte's own time and whether his positivism has ever given any useful advice about the sciences (1877, 1–3, 103–104, 119; 1878, 51–53). For instance, Comte's restrictions against hypotheses in probabilistic form undermined the larger positivist program of providing for a scientific re-organization of society, since so much of the practical advice science has to offer is expressed in probabilistic form (1897, 228).

Renouvier maintained that although the three-stage law was supposed to be a law of progress, Comte never provided a definition and a philosophical defense of what he meant by progress. Indeed, Renouvier thought that this was impossible for Comte, as there is no scientific or empirical way to define progress, which must be defined relative to some goal or ideal. For Comte to have defended such an ideal, he would have had to have embraced moral concepts that were not grounded in empirical science and that he had relegated to the metaphysical or theological stages (1867, 42, 46; 1875, 65–66; 1881a, 343; 1881b, 1, 3). According to Renouvier, positivists such as Comte who attempt to make morality a positive science confuse that which is with that which ought to be (1873b, 330). Normative questions could not be settled through empirical methods alone. Hence, even if the sciences could achieve certainty, which he doubted, the sciences alone do not provide an adequate basis for social policy.

The same held true for naturalist approaches to epistemology for Renouvier. For instance, he maintained that the question of the limits of human knowledge was a genuine philosophical problem that could not be solved through empirical methods alone (1873b, 324). Indeed, the very concept of knowledge for Renouvier entailed that of justification, which is an evaluative concept. Whether some empirical knowledge claim is justified is as much a normative issue as the justification of propositions in ethics and depends on the same conditions, in particular, on the possibility of freely deliberating and assenting to it.

Renouvier recognized that since science and indeed ethics depends on free deliberation, a political philosophy must grant people liberty of discussion. Yet the political philosophies of neither Comte nor Saint-Simon could guarantee the freedom to discuss diverse points of view. On the contrary, these philosophies sought homogeneity of thought. Comte had an uncritical confidence in the ability of scientific methods to answer social and political questions and assumed that these methods had reached their final stage of development in the *Cours*. Renouvier, on the other hand, believed that methods in the sciences were tentative in the same way that theories, hypotheses, and concepts were and would continue to change with time. He did not think that philosophers should legislate to scientists regarding the methods they should follow or the theories they should accept, but left it up to the scientific community to make their own decisions. He saw the philosopher's task instead as characterizing the process through which scientists reach agreement about such things. Thus his was a political philosophy of science in the sense **Pn**.

Scientists should enjoy the freedom to pursue even those lines of inquiry that he questioned, Renouvier thought. He raised empirical objections to Darwinian evolutionary theory, questioned Cantor's definition of infinite sets, and maintained that Lobachevsky's non-Euclidean geometry contradicted what he took to be the regulative principles of the understanding (1891, 43, 47, 64; 1889, 347). Nevertheless, he never prohibited research in these areas, the way that Comte had proscribed research in stellar astronomy as useless and the mathematical theories of probability and statistics as incompatible with the determinism of the positive philosophy (Renouvier 1901, 396). On the contrary, for instance, Renouvier granted mathematicians the liberty to pursue non-Euclidean geometries out of pure curiosity with the hope that they may one day yield useful results for science (1889, 346).

By the 1880s, Renouvier had adopted the term "convention" from mathematics and gave it a new meaning in the philosophy of science, in which he regarded methods and hypotheses as conventional, including the very hypotheses that constituted the subject matter of a science. The pursuit of inquiry in the sciences depended on a community of researchers agreeing for the time being not to challenge certain postulates, concepts, and presuppositions. Such an agreement constituted a social contract for Renouvier (1885, 2–3, 6–8). In the sciences, Renouvier argued, "there is only one foundation of certainty . . . and that is the consent and the agreement" of the scientists who participate in the development of the field (1885, 6). The existence of a science depends upon only two conditions: that it has an appropriate method to maintain and develop the science "and the constant adhesion of those who have accepted its initial positions and who maintain among themselves the convention of not submitting them to examination" (1885, 7). Among the things that scientists have agreed not to question, according to Renouvier, are Euclidean space in geometry, and the concepts of matter, cause, affinity, attraction, and repulsion in physics and chemistry. Other questions scientists have agreed not to raise are more metaphysical or epistemological, such as whether space and time exist outside our representations of them, whether quantities and indeed matter are continuous in themselves or only in our thoughts, and whether there exist souls interacting with bodies (1885, 7–8).

Renouvier's new use of the term "convention" in the philosophy of science was preceded by an earlier use of the term "social contract" to express a similar idea. For instance, he proposed that philosophy could become a science only through adopting "a sort of social contract" that would regulate discussion among them (1873c, 198). For Renouvier, these conventions or contracts, whether those governing a group of scientists engaged in a common pursuit or those governing societies at large, are living contracts, constantly modifiable through free and open debate. They begin as implicit or tacit agreements and are gradually made explicit over time (Renouvier 2002, 59–60; cf. 1912b, II, 193; 1912c, xxxix–xl). He used the concept of a social contract in order to answer a question that he thought Comte had failed to address: that is, what is the source or basis for normativity? Why should a scientist follow certain prescribed rules of method or accept certain principles in her discipline as given? For Renouvier, such questions were analogous to asking why people should follow legal and moral rules and received the same sort of answer:

one is obligated to follow legal, moral, methodological, or any other sort of rule when it is a part of a social contract to which one has voluntarily given one's consent (Schmaus 2011).

7.5 Summing Up

In all, I have distinguished at least fourteen possible meanings of the term political philosophy of science. Only in some of these senses does a political philosophy of science depend on the problematic underdetermination argument. One could argue that all of these senses, with perhaps the sole exception of **Cn2**, can be exemplified in twentieth century thought. But by looking to nineteenth century France, we have seen not only an example of **Cn2** and its critique, but alternative meta-ethical approaches to a political philosophy of science. By comparing the non-cognitivism of the logical positivists with Comte's naturalism and Renouvier's non-naturalism, we can get some insight into what possible avenues to pursue and which to avoid if we are interested in constructing a normative political philosophy of science. Of course, neither the nineteenth nor the twentieth centuries can provide exact models for a political philosophy of science today, as society has changed in many ways. But a comparison among these three approaches may help us clarify our thinking about these matters.

As we have seen, the link between philosophy of science and politics is a rather weak one for non-cognitivists. Non-cognitivism is not inconsistent with maintaining a political stance: one could be an emotivist about the political position one values. A politically committed emotivist could even see philosophy of science as serving her ideology in some way, such as in providing tools for clear thought. However, an emotivist could not mount an argument for her political preferences from her philosophy of science, or offer any other epistemic basis for her preferences. It is then hard to see how a non-cognitivist could defend a normative political philosophy of science.

Comte's naturalist philosophy would appear to exemplify Gomez's argument that pseudo-rationalism together with the assumption that political decisions should be based on science leads to authoritarianism. But what leads to authoritarianism is not the idea that decision making in science is value free, but the idea that the sciences alone suffice to provide correct answers regarding what is good for people, and that the sciences are already in possession of those correct ideas. Renouvier was thus opposed to Comte's idea of a scientific elite determining how we should live. For similar reasons, Renouvier was just as hostile to Catholicism as he was to political ideologies that pretended to be scientific. Both maintain that there are objective truths about right and wrong grounded in the nature of humankind, and appeal to an elite authority to tell us what those truths may be.

Whereas Comte's philosophy of science was political in the sense of providing the materials to construct a new political ideology (**Cn2**), Renouvier's was political in the sense of undermining this ideology by criticizing the underlying philosophy of science (**Cn1**). Renouvier's was also a political philosophy of science in the sense of characterizing and endorsing a social contract approach to the governance of science (**Pd**, **Pn**). His non-naturalism made room for the free and open discussion of normative questions and at the same time explained what gives the decisions reached their normative force.

As we mentioned above, we can find more recent examples of **Pd** and **Pn** in the works of Kitcher and Longino. Kitcher, at least, also explicitly rejects the philosophy that values have an objective basis in human nature (2001, 162ff). However, where Kitcher and Longino tend to draw their political philosophy from John Stuart Mill and consequentialism, Renouvier, although he did not ignore consequences, contributed a different perspective on the relationship between science and politics through his social contract approach. For instance, Kitcher affirms that our duty to pursue the truth, even into sensitive topics such as gender and racial differences, does not override our duty not to harm those who are less fortunate (Kitcher 2001, 103). But what is the basis of this duty? It may not be impossible for a consequentialist to explain duties, but Renouvier's social contract approach, which was grounded in a Kantian concern for the autonomy of the individual, could provide a direct answer to this question. For Renouvier, the social contract spells out our duties to one another, and a contract to which one has given one's voluntary consent, unlike a rule imposed by force, has moral authority. To say that Renouvier's social contract theory could provide an account of our duties to the less fortunate is not to suggest that Renouvier shared all of our contemporary concerns about race and gender. Nevertheless, it may be useful to consider a broad spectrum of political philosophies of science in addressing our current concerns.

In conclusion, for those interested in constructing a normative political philosophy of science today, non-cognitivism appears to be a non-starter. Naturalism is a possibility, as long as it is coupled with humility about our ability to achieve the truth about what is good. Given the limitations of human knowledge, people must be left free to debate and reach consensus on questions of values. But then naturalism becomes indistinguishable for all practical purposes from a non-naturalism such as Renouvier's, which grounds normativity in freely-given consent. A normative political philosophy of science, then, should provide for freedom of discussion both within the scientific community and in society at large.

Acknowledgements I wrote a dissertation on Comte under Peter many years ago. Thanks, Peter! The present section on Comte draws on an anthology of papers on Comte that we are currently revising for publication. The contributions by my co-editors Michel Bourdeau and Mary Pickering were especially helpful.

References

Bourdeau, Michel. 2006. *Les trois états: science, théologie et métaphysique chez Auguste Comte*. Paris: Éditions du Cerf.

Comte, Auguste. 1830–1842. *Cours de philosophie positive,* ed. Michel Serres et al. 2 Vols. Paris: Hermann, 1975.

———. 1844. *Discours sur l'esprit positif.* Paris: Carilian-Goeury et Victor Dalmont, Éditeurs.
———. 1851–1854. *Système de politique positive,* 4 Vols. Paris: Au siège de la Société positiviste, 1929.
———. 1875–1877. *System of Positive Polity,* 4 Vols. Trans. J. H. Bridges, et al. London. Reprinted New York: Burt Franklin, 1966.
———. 1988. *Introduction to Positive Philosophy*. Trans. Frederick Ferré. Indianapolis: Hackett Publishing Company.
Fedi, Laurent, and Raymond Huard. 2002. Avant-Propos, Presentation, and Other Front Matter. In ed. Renouvier, 2002, 5–33.
Feyerabend, Paul. 1987. *Farewell to Reason*. London/New York: Verso.
Gomez, Ambrosio Velasco. 2004. Toward a Political Philosophy of Science. *Philosophy Today* 48(5 supplement): 116–121.
Kitcher, Philip. 2001. *Science, Truth, and Democracy*. Oxford: Oxford University Press.
Kourany, Janet. 2010. *Philosophy of Science After Feminism*. Oxford/New York: Oxford University Press.
Longino, Helen. 1990. *Science as Social Knowledge*. Princeton: Princeton University Press.
———. 2002. *The Fate of Knowledge*. Princeton: Princeton University Press.
Pickering, Mary. 1993. *Auguste Comte: An Intellectual Biography*. Vol. I. Cambridge, MA: Cambridge University Press.
Popper, Karl R. 1965. *Conjectures and Refutations: The Growth of Scientific Knowledge*. 2nd ed. New York: Harper Torchbooks.
Renouvier, Charles Bernard. 1867. De la philosophie du XIXe siècle en France. *L'Année philosophique* 1: 1–108.
———. 1873a. La Prétensions de la Science. *La Critique philosophique* 2(1): 227–235.
———. 1873b. Review of *Principes de logique positive, ou Traité de scepticisme positif,* par F Saleta. *La Critique philosophique* 2(1): 323–323.
———. 1873c. Proposition d'un nouveau critère de certitude. *La Critique philosophique* 2(2): 195–201.
———. 1875. Y a-t-il une loi du progrès? *La Critique philosophique* 4(1): 65–68.
———. 1876. De la Méthode scientifique. *La Critique philosophique* 4(2): 401–404.
———. 1877. Le *Cours de philosophie positive* est-il au courant de la science? *La Critique philosophique* 6(1): 291–299, 327–336; 6(2): 1–7, 97–106, 113–20.
———. 1878. La Question de la Certitude. I – Le positivisme. *La Critique philosophique* 7(1): 49–53.
———. 1881a. Politique et socialisme. X. La Question du progrès. Burdin, Saint-Simon et Comte: La Théorie des époques organiques. *La Critique philosophique* 10(1): 336–346.
———. 1881b. Politique et socialisme. XI. La Question du progrès. Burdin, Saint-Simon et Comte: Le But du progrès. *La Critique philosophique* 10(2): 1–10.
———. 1885. Philosophie, science, et criticisme. *La Critique philosophique*, new series, 1st year, no. 1: 1–16.
———. 1889. La Philosophie de la règle et du compas, ou des jugements synthétiques a priori dans la géométrie élémentaire. *La Critique philosophique*, new series, 5th year, no. 2: 337–348.
———. 1891. La Philosophie de la règle et du compas. Théorie logique du jugement dans ses applications aux idées géométriques et à la méthode des géomètres. *L'Année philosophique* 2: 1–66.
———. 1897. *Philosophie analytique de l'histoire*. Vol. IV. Paris: Ernest Leroux.
———. 1901. *Histoire et solution des problèmes métaphysiques*. Paris: Félix Alcan, Éditeur.
———. 1912a. *Essais de Critique générale. Premier Essai. Traité de logique générale et de logique formelle*. 2 vols. Paris: Armand Colin.
———. 1912b. *Essais de Critique générale. Deuxième Essai. Traité de psychologie d'après les principes du criticisme*. 2 vols. Paris: Armand Colin.

———. 1912c. *Essais de Critique générale. Troisième Essai. Les Principes de la nature*. Paris: Armand Colin.
———. 2002. Sur le people, l'église et la république: Articles de 1850–1851. Suivi du Mémento rétrouvé. Ed. by Laurent Fedi et al. Paris: L'Harmattan.
Richardson, Sarah S. 2009a. The Left Vienna Circle, Part 1. Carnap, Neurath, and the Left Vienna Circle thesis. *Studies in History and Philosophy of Science* 40: 14–24.
———. 2009b. The Left Vienna Circle, Part 2. The Left Vienna Circle, Disciplinary History, and Feminist Philosophy of Science. *Studies in History and Philosophy of Science* 40: 167–174.
Rouse, Joseph. 1987. *Knowledge and Power: Toward a Political Philosophy of Science*. Ithaca: Cornell University Press.
Schmaus, Warren. 2011. Science and the Social Contract in Renouvier. *HOPOS: The Journal of the International Society for the History of Philosophy of Science* 1(1): 73–100. Republished in the Pantaneto Forum, issue 48, Oct. 2012, http://www.pantaneto.co.uk/issue48/schmaus.htm.
Shapin, Steven, and Simon Schaffer. 1985. *Leviathan and the Air-Pump*. Princeton: Princeton University Press.
Tresch, John. 2010. The Order of the Prophets: Series in Early French Social Science and Socialism. *History of Science* 48: 315–342.
Uebel, Thomas. 2005. Political Philosophy of Science in Logical Empiricism: The Left Vienna Circle. *Studies in History and Philosophy of Science* 36: 754–773.
———. 2010. What's Right About Carnap, Neurath and the Left Vienna Circle Thesis: A Refutation. *Studies in History and Philosophy of Science* 41: 214–221.

Chapter 8
Physicalism, Introspection, and Psychophysics: The Carnap/Duncker Exchange

Uljana Feest

Abstract In 1932, Rudolf Carnap published his article "Psychology in a Physical Language." The article prompted a critical response by the Gestalt psychologist Karl Duncker. The exchange is marked by mutual lack of comprehension. In this paper I will provide a contextualized explication of the exchange. I will show that Carnap's physicalism was deeply rooted in the psychophysical tradition that also informed Gestalt psychological research. By failing to acknowledge this, Carnap missed out on the possibility to enter into a serious debate and to forge an alliance with a like-minded psychologist at the time. I conclude by suggesting that the kind of physicalism practiced by Gestalt psychologists deserves to be taken seriously by current philosophy of psychology.

In the early 1930s, Rudolf Carnap laid down his project of overcoming metaphysics by means of linguistic analysis (Carnap 1931a) and specified a universal (physical) language as the language of choice (Carnap 1931b). It is well known that Carnap's 1931b article gave rise to what is often referred to as the "protocol-language debate" within the Vienna Circle (Neurath 1932; Carnap 1932b). While there is some impressive historical and philosophical scholarship about this debate (e.g., Uebel 2007), one strand of it has not received much attention, namely the ways in which Carnap's views about the physicalizability of protocol sentences were related to research in experimental psychology at the time. This is especially surprising given the fact that Carnap, in his article "Psychology in a Physical Language" (1932a) attempted to spell out the implications of this view for psychology. This article was met by a critical response by the psychologist Karl Duncker (1932), which in turn prompted a reply from Carnap (1932b). The exchange is characterized by a surprising degree of mutual incomprehension, with Duncker suggesting that Carnap's

This article is dedicated to Peter Machamer, who helped me find my voice as a historian and philosopher of science, and who taught me so much more than I realized at the time.

U. Feest (✉)
Institute of Philosophy, Leibniz University of Hannover, Hannover, Germany
e-mail: feest@philos.uni-hannover.de

M.P. Adams et al. (eds.), *Eppur si muove: Doing History and Philosophy of Science with Peter Machamer*, The Western Ontario Series in Philosophy of Science 81, DOI 10.1007/978-3-319-52768-0_8

critique of (introspective) psychology was attacking a strawman and Carnap saying that Duncker had completely missed his point.

In this article I will explicate and contextualize the exchange between Carnap and Duncker. I will shed some light on the reasons why the two talked past each other and I will show that Duncker did put his finger on the fact that (1) Carnap's position failed to address scientific practice, and that (2) Carnap did indeed attack several strawmen. I will lay out that Carnap's turn to a physical language was motivated by his aim to provide an objective foundation for protocol sentences (Sect. 8.1), and argue that the way in which Carnap executed his project of physicalizing protocol-sentences was deeply informed by psychophysics (Sect. 8.2). I will then (in Sect. 8.3) turn to Carnap's 1932a article "Psychology in a Physical Language," where he claims to be addressing the implications of his views for psychology. Section 8.4 details Duncker's response and Carnap's answer. Finally, in Sect. 8.5, I will draw out two underlying issues in this debate, i.e., (a) the status of introspection in psychological research, and (b) the question what (if any) metaphysical presuppositions were made by psychophysical research.

8.1 Overcoming Metaphysics and the Problem of Experience

In his famous article "Overcoming Metaphysics" Carnap laid out the very lean conception of philosophy characteristic of the Vienna Circle (Carnap 1931a). According to it, philosophy was neither to engage in metaphysical speculations about age-old topics, nor in naturalistic treatments of them. Rather, it was essentially reduced to providing meta-analyses of existing discourses in order to clean them of "pseudo-sentences" sentences that look grammatically like sentences, but are in fact meaningless. The method of choice (logical analysis of language) was to proceed by translating every sentence that is formulated in the so-called "material mode of speech" into a sentence in the "formal mode of speech" (a sentence about a sentence). This method was to reveal whether a given statement was logically consistent and empirically meaningful.

In response to the question of what it takes for a statement to be empirically meaningful, Carnap introduced a version of the well-known verificationist criterion of meaning that we still associate closely with the doctrine of logical positivism. According to it a word ***a*** is meaningful only if (1) empirical indicators for a are known, (2) it is known what protocol sentences the sentence S(***a***) can be derived from, and (3) the path towards verifying S(***a***) is known (Carnap 1931a, 224). Carnap's verificationist semantics for words emphasized the empirical truth conditions for sentences in which the words occur. These empirical truth conditions were provided by "observational" or "protocol-sentences" and he stated that while there was to date no agreement about the form or content of such sentences, they were commonly thought to refer to something that is "given" (Carnap 1931a, 222).

This raised the question of what were criteria of meaningfulness of protocol sentences themselves. Was their meaningfulness ensured by a primitive notion of the "given," i.e., by the subjective experience that is – in the material mode – reported by protocol sentences? Or was there a more 'public' way of stating the truth conditions for protocol sentences? It is this question that Carnap addressed in his "Physical Language as the Universal Language of Science" (Carnap 1931b), where he argued that both protocol sentences and "system sentences" (i.e., sentences capable of being derived from, and verified, by protocol sentences) are part of an overarching language: the universal language of science. By the requirement of *universality*, Carnap meant that such a language "can describe every state of affairs" (Carnap 1963 [1932a] 400). But in addition he argued that such a language should also be *intersubjective,* i.e., it should be usable by everybody in the same way. It is in this second respect that Carnap's aim in this work differed from the one proposed in his 1928 *Aufbau*, where he had wanted the universal language to be that of subjective experience. It was because of the requirement of *intersubjectivity* that Carnap turned to the physical language as the universal language (see Uebel 2007).[1]

Carnap's thesis that (seemingly subjective) protocol sentences were translatable into the (intersubjective) language of physics was ostensibly part of a larger thesis, namely that *all* sentences are translatable into sentences of the physical language. Surprisingly, Carnap did not present an argument for this larger thesis, except to suggest that its truth was obvious, at least in the case of "the inorganic sciences" (chemistry, geology, astronomy) and even biology since they were (in the material mode) dealing with physical objects. However, since his main target was the physicalizability of protocol sentences, a separate argument was required, and he attempted to provide one in Sect. 8.3 of his "Physical Language."

8.2 The Psychophysical Underpinnings of Physicalized Protocol Sentences

Carnap's argument for the translatability of protocol sentences into physical sentences took the form of an empirical claim: He posited that it is possible to find a quantitative equivalent for every qualitative (protocol) sentence, and he noted that this is not a logical necessity but simply a contingent empirical fact, such that there is a functional dependency between protocol sentences and physical sentences. (In the material mode, it is an empirical fact about the way in which our experience is structured in relation to the physical world). From this, Carnap derived the claim that it is possible to find a corresponding physical sentence for every protocol sentence, a fact that provides the basis of the very possibility of practicing an intersubjective physics.

[1] In his "Physical Language" article he still maintained that protocol-sentences were the most basic sentences of science that could not themselves be doubted (438), but in response to Neurath's critique, he revised this position to say that any scientific sentence within a physicalist system could function as a protocol sentence ("Über Protokollsätze" 224) and no sentence could function as an ultimate epistemic basis (225).

It is clear that Carnap expected the physical sentences in question to be sentences about the brain, but realized that neuroscience at the time did not deliver sentences that directly corresponded to sentences about experiences. For this reason, he decided to describe the relevant brain states in terms of behavioral dispositions, specifically, the disposition to display particular behaviors in response to particular stimuli. This was made explicit in his subsequent "Psychology in a Physical Language," but is already apparent in the way he explains his position in his 1932b paper. For example, he states that it is possible to establish empirical correlations between the qualitative statements of protocol sentences and the quantitative determination (in a physical language) of the conditions under which they are uttered. For example, when examining color vision, he argued, one has to vary "the physical conditions (e.g., the combination of various frequencies of oscillations) and discover the conditions to which S reacts with the protocol statements containing the qualitative terms in question" (Carnap 1963 [1932]; 408). Then he states "The discovery of the set of these physical terms corresponding to a definite qualitative term will be called the 'physicalizing" … of this qualitative term" (ibid.).[2]

Unfortunately, Carnap does not provide a reference for this, but it is striking that there was in fact a research program that did just what Carnap was describing: i.e., vary physical stimulus conditions and measure responses. This research program, psychophysics, had famously been founded by Fechner (1860) and continues to be active until today (Heidelberger 2004a). At the time of Carnap's work in the 1930s, famous proponents of this research were members of the Berlin/Frankfurt school of Gestalt psychology, with whom Carnap was at least indirectly acquainted (Feest 2007). I therefore suggest that Carnap was aware of psychophysical research and that his proposal to translate protocol sentences into sentences about behavioral dispositions was in fact inspired by this tradition.[3] Carnap's thesis of the physicalizability of experiential statements, thus relied on a contingent empirical fact. And the way in which he imagined the empirical investigation of this fact was practiced by a particular (at the time quite dominant) approach within psychology. The implication I want to highlight here is that specific attempts to translate a psychological sentence (Mr A is angry) into physical sentence (Mr. A exhibits particular behaviors) are going to build on psychophysical research, which in turn will necessarily involve first-person data.

[2] It follows that even though Carnap took every experiential sentence to be extensionally equivalent with a physical sentence, this equivalence was, for him, an empirical fact, not an analytical truth. Crawford (2014) therefore argues that Carnap's position is not accurately described as "analytical behaviorism."

[3] Even if this historical thesis does not hold up, I maintain that psychologists at the time would have recognized the similarity (and that's all that matters for my subsequent argument).

8.3 Physical Language, Physicalized Observation Sentences and Psychology

In his article "Psychology in a Physical Language" Carnap explored the implications of his physicalism for psychology as a science, specifically focusing on the question of the kinds of observation sentences admissible in psychology. His targets were "observations of the mental states of others" (Sect. 8.2) and "self-observations" (Sect. 8.6).

With regard to observation-statements about the mental states of others (e.g., "Mr A is angry"), Carnap argued that insofar as such sentences are meaningful at all, they are only meaningful if they can be translated into statements about physical behaviors (about Mr. A's disposition to behave in certain ways). This allows for the derivation of sentences that state truth conditions for the sentence in question (about Mr. A's actual behavior), though (as Carnap lays out) to treat them as such requires an additional premise, namely that in general people display the behaviors in question when angry. Carnap uses this to argue that if we want to test a particular sentence about the content of someone's mind, we have to (a) appeal to a general sentence about the kinds of physical conditions that need to be in place when we use the term "anger" and (b) point to a particular set of physical conditions as instantiating the general conditions in question. Carnap then contrasts this ("rational") mode of justification with one where the emotional state of Mr. A is ascertained in a more "intuitive" way. He argues, however, that such intuitive sentences are either meaningless or can be translated into one that states the physical conditions that provide it with meaning. In Sect. 8.6, he picks up on this claim and says that the same is true of sentences that we might utter about our own mental states: For example, when we utter a sentence like "I am nervous right now," this sentence is either meaningless or its meaning is provided by empirical truth conditions in a physical language (shaky hands, sweaty palms, etc.).

One might wonder whether (and if so, how) these considerations were relevant to the research practices of experimental psychology, as Carnap seems to suggest. In a nutshell, he had two answers to this: First, he claimed that by failing to appreciate his point about the semantics of psychological sentences, psychologists were prone to falling into a kind of psychophysical dualism (which is apparent, for example, when I say that I am in physical state X and *in addition* in mental state Y). Second, he cautioned against attributing a special kind of epistemic significance to first-person experiences (of other minds or of one's own mind), pointing out instead that in science one always deals with *sentences* about experiences, which should be treated as the behavioral outputs of complicated detection devices under particular physical conditions: "In principle, there is at most a gradual epistemic difference between the utterances of a fellow human being and a barometer" (p.124; translation by me). (see also p. 140 for similar statements).

Carnap repeatedly comments on the confused state of the then current psychology (for example "understanding" and "introspective" psychology), but he never quite says who he is actually accusing of committing the above two errors. While not mentioning him by name, it is quite possible that he had in mind Wilhelm Dilthey, who had argued for understanding as a first-person mode of access to the subject matter of the human sciences. But Carnap's critique is a little confusing by virtue of the fact that Carnap also throws in a different psychological approach (again without mentioning any names), which studies "purposeful behaviour." Here, Carnap may have had in mind the American neo-behaviorist Edward Chace Tolman, whom Carnap probably met at Vienna Circle meetings (or at least knew about via Egon Brunswick).[4] Tolman emphasized the goal-directed nature of much behavior and essentially introduced proto-cognitive mental states to explain them. With respect to this type of explanatory practice, Carnap argues that it is acceptable as long as we realize that talk about purposes can be fully physicalized, i.e., translated into a language that specifies lawful regularities between specific stimuli and behavioural dispositions. It bears stressing that Dilthey's approach was quite different from Tolman's, insofar as the former talked about a mode of empirical access (understanding), whereas the latter talked about an explanatory concept (purpose). Hence, we can note that Carnap's critique of psychology was fairly broad und unspecific.

Although this does not come out in Carnap's article, a more immediate target of Carnap's attack on psychology was probably the psychologist Karl Bühler, who was based in Vienna at the time. In 1926 and 1927, just a few years before Carnap's "Psychology in a Physical Language", Bühler had published a long paper (Bühler 1926) and a book (Bühler 1927), both of which concerned themselves with the "crisis of psychology", a topic that was being discussed by various authors at the time (cf. Sturm/Mühlberger 2012). As Sturm (2012) points out, Bühler took a moderate position with respect to this "crisis." He agreed that the methodological and theoretical fragmentation of the discipline presented problems, but he did not advocate for any one approach as presenting a solution to these problems. Instead, he suggested that it was necessary and desirable to combine three methodological approaches popular at the time: One that focused on the phenomenological description of subjective experience (as exemplified, for example, in the Würzburg School of thought psychology), one that focused on purposeful behavior (as rejected by classical behaviorism), and one that focused on the formation of the objective mind (as suggested, for example, by Wilhelm Dilthey). Bühler did, however, strongly oppose the idea that "physicalism" could be regarded as presenting a viable methodological and theoretical cure for the problems that beset the discipline of psychology, and he specifically mentioned the position of Gestalt psychologists like Wertheimer, Köhler, and Koffka as exemplifying the kind of physicalism he rejected.

I have argued elsewhere that members of the Berlin/Frankfurt school of Gestalt psychology should be classified as having practiced (a version of) phenomenological methods, albeit combined with a psychophysical paradigm (cf. Feest 2014). That aside, it is also clear that Wertheimer and his colleagues did aim to firmly tie subjective experiences to specific and quantifiable stimulus configurations, hence in a sense physicalizing them. This interpretation of their work comes out especially

[4] Tolman's book, *Purposive Behavior in Animals and Men* also appeared in 1932.

clearly in Schlick's 2008 *Allgemeine Erkenntnislehre*, where he rejected "introspective psychology" and strongly endorsed the project of "physicalizing" qualitative mental states, singling out Gestalt psychologists of the Berlin/Frankfurt school (see also Wendel and Engler 2008). According to Christoph Limbeck (2014), Carnap presented his ideas about a physicalist psychology in Bühler's colloquium on two separate occasions in the summer of 1930, where they gave rise to heated discussions.[5] It is thus not too far-fetched to speculate that Carnap's critique of contemporary psychology may have been directed (at least indirectly) at the ideas expressed in Bühler's 1927 book, which are likely to have informed Bühler's reaction to Carnap's approach. In the light of the fact that Bühler, in this book, specifically opposed what he took to be the "physicalism" of the Berlin school of Gestalt psychology, it is not surprising that members of this school took an interest in Carnap's defense of a "psychology in a physical language." (Carnap 1932a, 1932b)

8.4 The Carnap/Duncker Exchange

Carnap's 1932a article prompted a reply by Karl Duncker, a younger member of the Berlin school of Gestalt psychologists. This reply (and Carnap's subsequent response) demonstrates a deep mutual incomprehension. This is especially surprising in the light of my above thesis that Carnap's physicalism was informed by the psychophysical tradition, and that Gestalt psychological research has to be placed in that tradition. In this section, I provide a brief overview of the exchange, followed (in Sect. 8.5) by an elaboration of my thesis.

Even though it is clear, especially at the beginning of his article, that Duncker had misunderstood some of Carnap's points, we should take seriously his objection to Carnap's characterization of the two supposed problems of psychology, namely the danger of falling into a mind-body dualism and the tendency to attribute too much epistemic significance to introspective data. Carnap had argued that these two problems could be avoided if one took the general stance of behaviorism: "The position advocated here is essentially in agreement with the psychology known as 'behaviorism'" (Carnap 1932a, 124, translation by me).[6] Duncker objected that Gestalt psychological findings and methodology were much more congenial to what Carnap was aiming at.

As mentioned above, Carnap claimed that his analysis of psychological sentences as translatable into physical ones could help psychologist avoid the inference that the two types of sentences referred to two separate kinds of "things." Duncker responded with utter incomprehension, stating that he was not aware of many contemporary psychologists who made substantial metaphysical assumptions about

[5] I would like to thank Christoph Limbeck for drawing my attention to this! (As Sturm, 2012, points out, this has also been noted by Toccafondi, 2004).

[6] Here Carnap mentions that J. B. Watson's *Der Behaviorismus* had been translated into German in 1930.

mind and body (Duncker 1932, 165), and that Gestalt psychologists were in fact physicalists, though he did not think that this had any implications for the goals or methods of non-behaviorist psychology (ibid. 176). He backed this up by claiming that psychophysics also viewed organisms as detection devices, like Carnap, and that they endorsed a principle of isomorphism between mental and physical processes (Duncker 1932, 174). However, Duncker also used some careless formulations, which seemingly contradicted his anti-metaphysical stance. For example, he asserted that it was possible to conceptualize anger both as a behavioral disposition and as an inner experience. For Carnap these assertions showed Duncker to be falling in exactly the metaphysical trap that Carnap had warned about (Carnap 1932b, 186/7). Accordingly, Carnap responded with some surprise about Dunker's claim to be a physicalist, noting that he and Duncker clearly had in mind very different notions of *physicalism*: Carnap's physicalism was about the translatability of psychological language into physical language, whereas Duncker's physicalism was about finding the physical basis of introspectively accessible experience.

This brings us to Carnap's second critique of psychology, namely that of introspective methods. Carnap's rejection of introspection was closely related to his above-mentioned point about the dangers of being misled into a dualist metaphysical position. It was not aimed at the usage of this method as such, but at certain interpretations of its results, with its inherent danger of reifying the object of introspection, a point he reiterates in his reply to Duncker. In his response, Duncker stated that (a) Gestalt psychology (like all of psychophysics) relied on some kinds of self-reports since it was not clear how psychophysical laws could be formulated otherwise, but (b) that one did not have to be a behaviorist to reject the assumption that psychology aims at the "absolute content of a quale" (Duncker 1932, 167). Moreover, (c) when psychophysicists investigated (for example) color vision, they were not interested in the subjective experience of the color as such, but in their "order properties" ("Ordnungseigenschaften"). For example, if an individual experienced inverted qualia, this was irrelevant for the language of science, so long as that person responded to all the same stimuli in the same way as a person with 'normal' experiences. All of these points then lead up to Dunckers's somewhat exasperated question: "I ask once again, which 'interpretation' of introspection is Carnap arguing against?" (Duncker 1932, 169).

Summing up, Duncker held that an endorsement of introspection was compatible with a lean metaphysics and with the notion of biological organisms as physical detection devices. Yet, at the same time, he held that introspection was an indispensable tool for the empirical investigation of such devices. Carnap, in his reply, repeated that he did not care what tools psychologists used, so long as the resulting sentences could be stated in a physical language. Thus, he took Duncker's conjectures to be irrelevant to his point.

8.5 Some Underlying Issues

In this section, I will argue that while it is true that Duncker misunderstood Carnap's point, it is also the case that Carnap was not receptive to the implications of Duncker's remarks, namely (a) that Carnap's physicalism did not have much practical relevance to at least some non-behaviorist psychology as it was being practiced at the time (or, to the extent that it did, Carnap had failed to engage with the targets of his critique), and (b) that Carnap's physicalism was rooted in psychological research and hence he would have been well advised to engage with the foundations of that research.

8.5.1 The Heuristic Function of Introspection

Clearly, one issue at play in the exchange between Carnap and Duncker was a version of the distinction between discovery and justification, with Carnap declaring scientific methodology irrelevant to the epistemological status of psychological claims such as "Mr A is angry." At that point in time, questions surrounding the production and epistemic status of first-person experiential reports had been debated within experimental psychology for well over half a century, following the work of Fechner, Brentano, G.E. Müller, and many others. The fact that Carnap simply ignored these debates must have seemed bizarre to a practicing scientist like Karl Duncker. Now, Carnap might have replied that those very discussions were themselves rooted in metaphysical assumptions, and that it was precisely for this reason that he was aiming at a purely formal analysis of scientific language rather than the metaphysical presuppositions of scientific practice. But the question is whether Carnap was entitled to this stance, since – as I argued above – his very conception of the physicalization of protocol sentence took for granted the project of psychophysics, that is, the project of discovering functional relationships between physicalist and mentalist descriptions. It is this fact that Duncker is also alluding to when arguing that even though experimental psychology mainly varies physical stimuli, this activity is heuristically guided by introspection (Duncker 1932, 173). This can be illustrated if we go back to Carnap's claim (see Sect. 8.3 above) that the empirical conditions of application for a psychological sentence (such as "Mr. A is angry") were provided not only by an individual displaying the relevant behaviors, but also by an empirical law that described the types of behaviors typically displayed by angry people. Such laws, Carnap tells us, are the result of inductive generalizations. What Duncker is pointing out (on my construal) is that such generalizations are based on self-reports, and that therefore human subjectivity cannot be eliminated from the research process.

Again, Carnap might have replied that it was precisely the (merely) heuristic nature of introspection that rendered it irrelevant to serious logical analysis, and moreover, that even in the actual research (as he and Duncker agreed) only the introspective *reports* (not the introspective experiences themselves) counted. But even if we grant this, I would like to suggest that some of the unproductive harshness of this exchange could have been avoided if Carnap had acknowledged his debts to the psychophysical tradition. This might have helped him understand why Duncker was so irritated by Carnap's positive assessment of methodological behaviorism, since after all behaviorism, by virtue of not talking about mental states at all (physicalized or not) radically rejected the very method that Carnap's physicalism was built on, namely that of treating experiential reports as relevant scientific data. I will now argue that it might also have helped him provide a more nuanced description of the kinds of mind-body parallelisms available at the time.

8.5.2 Varieties of Mind-Body Parallelisms

As indicated above, Duncker rejected Carnap's claim that contemporary psychology's use of the material mode was leading it down the road to mind-brain dualism. But what did Duncker have in mind here?

Carnap's turn to the analysis of language was motivated by his aim to avoid metaphysics. While the Vienna Circle's program of anti-metaphysics is particularly well known, the mid- to late nineteenth century had seen a lot of debates about banning metaphysics from scientific and philosophical discourses. In this vein, already Fechner, in his 1860 *Elements of Psychophysics*, had formulated an account of the psychophysical relationship that aimed to steer clear of fruitless debates between materialists and idealists at the time. His response was to argue that mental and physical properties were token identical, but depending on one's perspective, one could only ever empirically apprehend one or the other and never both at the same time (Fechner 1860; Heidelberger 2004a, b). While this type of position was often referred to as a kind of "parallelism" (see also Heidelberger 2003), a more apt description might be "dual-aspect theory," since this term does not suggest the existence of distinct substances or properties, but merely of distinct perspectives. It was precisely this notion that underwrote Fechner's empirical project of psychophysics. As Heidelberger (2003) explains, it is possible to distinguish between three layers of Fechner's parallelism: an empirical hypothesis about the functional relationship between physical and psychological descriptions, a dual-aspect theory of the mind/body relationship, and a cosmological thesis, according to which even inorganic processes have a mental side to them. I have argued above that Carnap not only shared Fechner's empirical hypothesis, but also his methodology of how to investigate this functional relationship (by varying stimuli and recording responses). While he clearly did not agree with Fechner's mind-body theory (let alone with his cosmological thesis), it bears stressing that Fechner's mind-body theory was not a kind of substance dualism. Rather, it was a dual aspect theory which scholars like Duncker may well have taken to lay the foundations for the very possibility of the psychophysical research that Carnap implicitly appealed to, when casting protocol sentences in terms of behavioural dispositions.

Now, it is clear that for Carnap, the linguistic description of behavioral dispositions (expressed in a physical language) merely ensured the meaningfulness of psychological sentences, whereas for Fechner they expressed psychophysical laws, i.e., laws that describe the empirical relationship between two types of the magnitudes: experiences and physical stimuli. But given that Carnap's semantic analysis also exploited (or at least assumed) the lawful empirical relationship in question, it is well pointing out that Fechnerian psychophysics attempted to account for the existence of the empirical relationship without positing separate substances or even properties. By stating that the correspondence between the two languages was a crude empirical fact, Carnap may have been able to avoid metaphysical speculation, but there is also a sense in which this appeal is somewhat unsatisfactory. Moreover, it remains to me an open question to what extent Carnap's thinking about this was implicitly informed by some version of the dual-aspect theory that had underwritten Fechner's psychophysical research. Whether or not this was the case, I argue that for Duncker this may have been an intuitive way to think about the matter, which would account for his difficulties in comprehending Carnap's point.

8.6 Conclusion

I have argued that Carnap's account of protocol sentences (including those of psychology) was deeply informed by the psychophysical research tradition of the mid-nineteenth to early twentieth centuries. In the light of this, I have provided an analysis of the exchange between Carnap and Duncker, in which Duncker questioned Carnap's contention that the methodological approach of behaviorism within psychology was congenial to his approach, arguing instead that Gestalt psychology came much closer to Carnap's outlook. I substantiated Duncker's assessment by providing a reading of Duncker's analysis that highlights the following two points: First, Gestalt-psychological research (and psychophysical research more generally), while giving a lot of weight to first-person experiential reports, did not necessarily invest them with epistemic certainty or treat them as being about irreducible qualia. Second, researchers in the psychophysical tradition (including Gestalt psychologists) were not necessarily committed to a mind-brain dualism, even if they aimed to formulate psycho-physical laws.

Given that Carnap's formal analysis relied on (or at least presupposed the possibility of) this research, I argue that it was unwise for Carnap to reject as irrelevant Duncker's points, both because it unnecessarily alienated a potential psychological ally and because it prevented Carnap from acknowledging the extent to which he and Duncker shared similar philosophical roots. It also prevented him from recognizing that his project of physicalizing protocol sentences (in the formal mode) relied on research that granted some epistemic authority to subjective experience (in the material mode).

In conclusion, I argue that the way in which Carnap tried to insulate his philosophical project, as concerned with the "epistemological status" of psychological sentences, from the question of how such sentences are established was part of a general trend away from being concerned with scientific practice. While philosophy of psychology in the nineteenth century had still been fairly practice-oriented (as evidenced, for example, in Fechner's 1860 *Elements of Psychophysics* or Brentano's 1874 *Psychology from an Empirical Standpoint*), the philosophical turn to formal analysis (along with the rise of behaviorism in the US), for some decades eclipsed the fact that much of psychology continued to make some use of first-person reports. As a result philosophers of psychology have only recently started to turn their renewed attention to questions about the meaning, role and justification of first-person reports in psychology's research practices (e.g., Hatfield 2014; Chirimuta 2014). In this context, the physicalist analyses of introspection, as they were provided by advocates of Gestalt psychology, are still well worth considering (Feest 2014).

Acknowledgements An earlier version of this paper was presented at the 2014 meeting of the *International Society for the History of Philosophy of Science* (HOPOS). I wish to thank members of the audience for a helpful discussion.

References

Bühler, Karl. 1926. Die Krise der Psychologie. *Kant Studien* 31(1–3): 455–526.

———. 1927. *Die Krise der Psychologie.* Jena: Verlag Gustav Fischer.

Carnap, Rudolf. 1931a. Überwindung der Metaphysik durch logische Analyse der Sprache. *Erkenntnis* 2: 219–241.

———. 1931b. Die physikalische Sprache als Universalsprache der Wissenschaft. *Erkenntnis* 2: 432–465.

———. 1932a. Psychologie in physikalischer Sprache. *Erkenntnis* 3: 107–142.

———. 1932b. Erwiderung auf die vorstehenden Aufsätze von E Zilsel und K. Duncker. *Erkenntnis* 3: 177–188.

———. 1963. The Physical Language as the Universal Language of Science. In *Readings in Twentieth-Century Philosophy*, ed. W. Alston and G. Nakhnikian, 393–424. New York: The Free Press [originally published as Carnap 1931b].

Chirimuuta, Mazviita. 2014. Psychophysical Methods and the Evasion of Introspection. *Philosophy of Science* 81(5): 914–926.

Crawford, Sean. 2014. On the Logical Positivists' Philosophy of Psychology: Laying a Legend to Rest. In *New Directions in the Philosophy of Science*, *The Philosophy of Science in a European Perspective*, ed. M.C. Galavotti et al., Vol. 5, 711–726.

Duncker, Karl. 1932. Behaviorismus und Gestaltpsychologie (Kritische Bemerkungen zu Carnaps ‚Psychologie in physikalischer Sprache'). Erkenntnis 3: 162–176

Feest, Uljana. 2014. Phenomenal Experiences, First-Person Methods, and the Artificiality of Experimental Data. *Philosophy of Science* 81: 927–939.

———. 2014. Phenomenal Experiences, First-Person Methods, and the Artificiality of Experimental Data. *Philosophy of Science* 81: 927–939.

Fechner, Gustav Theodor. 1860. *Elemente der Psychophysik.* Leipzig: Breitkopf und Härtel.

Hatfield, Gary. 2014. Psychological Experiments and Phenomenal Experience in Size and Shape Constancy. *Philosophy of Science* 81(5): 940–953.

Heidelberger, Michael. 2003. The Mind-Body Problem in the Origin of Logical Empiricism: Feigl and Psychophysical Parallelism. In *Logical Empiricism: Historical and Contemporary*

Perspectives, ed. P. Parrini, M. Salmon, and W. Salmon. Pittsburgh: Pittsburgh University Press.

———. 2004a. Fechner's (Wider) Conception of Psychophysics – Then and Now. Contribution to "Fechner Day 2004" XXth Meeting of the *International Society for Psychophysics*. Coimbra, Portugal, 18–22 October, 2004 (unpublished).

———. 2004b. *Nature From Within: Gustav Theodor Fechner and His Psychophysical Worldview*. Pittsburgh: University of Pittsburgh Press (Translated from the German by C. Klohr).

Limbeck, Christoph. 2014. *Der Physikalismus bei Bühler und Carnap* (unpublished manuscript).

Neurath, Otto. 1932. Protokollsätze. *Erkenntnis* 3: 204–214.

Schlick, Moritz. 2008. *Allgemeine Erkenntnislehre*. Moritz Schlick Gesamtausgabe Band I/1, ed. Wendel and Engler. Wien/New York: Springer [first published 1918].

Sturm, Thomas. 2012. Bühler and Popper: Kantian Therapies for the Crisis in Psychology. *Studies in History and Philosophy of Biological and Biomedical Sciences* 43(2): 462–472.

Sturm, Thomas, and Annette Mühlberger. 2012. Crisis Discussions in Psychology—New Historical and Philosophical Perspectives. *Studies in History and Philosophy of Biological and Biomedical Sciences* 43(2): 425–433.

Toccafondi, F. 2004. *Karl Bühler and neopositivism: Reasons for a non-convergence. Histoire, Epistémologie, Langage, 2004, Complément électronique N°2*. http://htl.linguist.jussieu.fr/dosHEL.htm

Uebel, Thomas. 2007. *Empiricism at the Crossroads. The Vienna Circle's Protocol-Sentence Debate*. Chicago: Open Court.

Wendel, Hans Jürgen and Engler, Fynn. 2008. "Einleitung". In Moritz Schlick. *Allgemeine Erkenntnislehre*. Moritz Schlick Gesamtausgabe Band I/1, ed. Wendel and Engler, 9–49. Wien/New York: Springer.

Chapter 9
Long-Term Potentiation: One Kind or Many?

Jacqueline A. Sullivan

Abstract Do neurobiologists aim to discover natural kinds? I address this question in this chapter via a critical analysis of classification practices operative across the 43-year history of research on long-term potentiation (LTP). I suggest that this 43-year history supports the idea that the structure of scientific practice surrounding LTP research has remained an obstacle to the discovery of natural kinds as philosophers of science have traditionally conceived them.

9.1 Introduction

A unique aspect of being trained by Peter Machamer is the importance he places on understanding scientific practices, providing descriptively accurate accounts of those practices and teasing out the interesting philosophical implications. He also encourages his students to gain hands-on experience in those areas of science of interest to them. My experience working in a neurobiology laboratory as a graduate student in combination with his mentoring have been invaluable to my thinking on a host of philosophical issues, including the one I take up in this chapter: the relationship between neurobiological kinds and scientific practice.

To provide some relevant background, in a paper in 2009, I described an observation I made while undertaking research in a neurobiological laboratory. I worked in the field of synaptic plasticity, the ability of synapses to undergo changes in response to patterns of electrical activity, which is thought to underlie learning and memory. I noticed that neuroscience affords investigators the freedom to produce forms of synaptic plasticity in a plurality of different ways, using different patterns of afferent stimulation. Given that different stimulation protocols could potentially recruit different mechanisms, it was an open question whether phenomena produced in different laboratories using different protocols were the same phenomenon or different phenomena.

J.A. Sullivan (✉)
Department of Philosophy, Rotman Institute of Philosophy,
University of Western Ontario, London, ON, Canada
e-mail: jsulli29@uwo.ca

M.P. Adams et al. (eds.), *Eppur si muove: Doing History and Philosophy of Science with Peter Machamer*, The Western Ontario Series in Philosophy of Science 81, DOI 10.1007/978-3-319-52768-0_9

One of the examples I used in my paper to illustrate the "multiplicity of experimental protocols" was long-term potentiation (LTP), which is generally defined as an activity-dependent increase in the strength of a synapse. I restricted my focus to then contemporary investigations of the role of a single protein kinase cascade in LTP, the extracellular-signal regulated kinase (ERK), and provided evidence that in response to different LTP induction protocols (e.g., theta-burst, high-frequency stimulation), all which investigators claimed could be used to successfully induce LTP, the kinase responded differently. I used this as a basis to suggest that it was a live possibility that different investigators who used different LTP induction protocols to investigate the mechanisms of LTP were actually investigating different phenomena produced by different mechanisms rather than the same phenomenon produced by the same or different mechanisms. In other words, given the structure of experimental practice, it was unclear if LTP was one kind or many kinds.

Some questions that I did not ask in the 2009 paper, however, were: (1) Do LTP researchers themselves take different instances of LTP to be the same kind of phenomenon or different phenomena? (2) Is there consensus in the field about how to "lump" or "split" the phenomena? (3) Is the multiplicity of experimental protocols an obstacle to the discovery of kinds that track actual divisions in the causal structure of the world (i.e., so-called "natural kinds")?

In this chapter, I engage in an historical analysis of LTP in order to provide at least some preliminary answers to these questions. I begin in Sect. 9.2 with an analysis of two strategies that have been put forward in the philosophical literature on mechanisms for revising scientific taxonomies, what I will refer to simply as (a) the "natural kinds strategy" (Craver 2002) and (b) the "conventional kinds" strategy (Craver 2009). In Sect. 9.3, I use these strategies in combination with some conceptual tools for thinking about LTP experiments in order to answer the aforementioned questions.

9.2 Two Competing Constraints on Scientific Taxonomies

It is widely accepted that a primary aim of neuroscience is to describe the mechanisms that produce phenomena of interest (e.g., Bechtel 2008; Machamer et al. 2000; Craver 2007). Implicit in this account of mechanistic explanation is a characterization of scientific progress in which kinds of phenomena and the mechanisms productive of those phenomena change over time. The basic idea is that sciences like neuroscience begin by characterizing phenomena and organizing them into groups on the basis of detectable surface features. As empirical inquiry yields data concerning the mechanisms that produce these phenomena, revisions are made with the aim of accommodating that taxonomy to the mechanistic structure of the world. This process is iterative; as more is found out about the mechanisms underlying phenomena of interest, further revisions to the taxonomy are possible (e.g., Bechtel 2008; Bechtel and Richardson 1993; Craver 2007, 2009). This view is consistent with the idea that sciences like neuroscience have realist aims; they aim to provide explanations of phenomena that reflect "how actually" those phenomena are produced in the natural world. In doing so, they approximate towards scientific taxonomies that reflect real divisions of kinds that correspond to the mechanistic structure of the world (See Craver 2006).

Areas of science that seek to accommodate their taxonomies of kinds to the mechanistic structure of the world may be described as being engaged in a search for so- called "natural kinds" (Craver 2002, 2007, 2009). In an early paper, Carl Craver (2002), noted that cognitive neuroscientists seemed to uphold a sufficient criterion for natural kindhood that he dubbed the "No Dissociable Realization (NDR)" condition.[1] On this condition, "instances of a natural kind have one and only one realizer" and "if there are two distinct realizers for a putative instance of a kind, there are really two kinds, one for each realizer" (Craver 2002, 962). The example that Craver provides to illustrate how this condition operates in practice is memory research. Although neuroscientists originally believed that memory was a single kind of phenomenon, findings from lesion studies and research on subjects with selective brain damage revealed that declarative memory depends on different brain structures than procedural memory. In light of these findings, the category of memory was split into two groups of phenomena. Further findings about memory mechanisms have prompted further subdivisions in the taxonomy of memory (e.g., Kandel and Squire 2008; Sweatt 2009). These and other examples of taxonomic practices in action (e.g., Bechtel 2008) lend support to the idea that cognitive neuroscientists implement the NDR condition when revising the taxonomy of kinds of memory. Thus, in instances where empirical inquiry yields findings that indicate that the current taxonomy of neurobiological kinds does not correspond to neural architecture, the taxonomy is revised so as to ensure such correspondence.

Craver (2009), however, offers an alternative account of the kinds of criteria that inform "taxonomic revisions" in neuroscience that is also compatible with mechanistic accounts of neuroscientific explanation. Whereas the NDR condition is consistent with the idea that neuroscientists treat the world as "an objective arbiter among competing taxonomies of kinds" (e.g., of memory) and that their explanatory interests play no role in shaping the kinds they ultimately discover, if we consider the situation from another perspective, claims Craver, we become privy to the fact that investigators themselves, rather than the world, determine when two kinds of phenomena or two kinds of mechanisms are different or the same. Consider again the declarative and procedural memory case. Empirical findings across animal and human studies indicate that declarative memory depends on structures in the medial temporal lobe and procedural memory depends on the basal ganglia. However, the first thing to notice is that the kinds of findings supporting this division emanate from different kinds of brains—animal and human brains—that differ from each other anatomically and mechanistically. Further, within a given human population and within the context of even a single experiment no two hippocampi or basal ganglia are alike—they vary in terms of a number of structural and constitutive details (e.g., overall shape of brain area, cell number, numbers of synaptic connec-

[1] I am using this condition that Craver puts forward in his 2002 paper as a heuristic because I think it gets something right about how some neuroscientists conceive of double dissociation experiments and what can be accomplished by using them.

tions, intracellular and extracellular molecular concentrations). Variations in behavioral performance and gross brain activity (e.g., BOLD signal) across subjects are also common in experiments in cognitive neuroscience.

The aforementioned differences could all be taken, from the perspective of an investigator, to correspond to different joints of nature. However, from the perspective of scientists, to acknowledge such differences as relevant for building a scientific classification system would result in an unwieldy taxonomy of kinds that would contain as many different kinds as there are different individuals (either animals or humans). This would make the taxonomy wildly intractable and bar the kind of generality that neuroscientists desire for their explanations. Investigators must instead strike a delicate balance between "characterizing [a] mechanism very abstractly", which "potentially glosses over sub-kinds of mechanism" and "characterizing [a] mechanism in maximal detail", which "threatens to make each particular mechanism a kind unto itself" (Craver 2009, 587; See also Craver 2014).

Striking such a balance is, in fact, what cognitive neuroscientists may be characterized as doing. The taxonomic division between procedural and declarative memory abstracts away from certain differences while acknowledging others. The upshot is that rather than adhering strictly to developing taxonomies to directly mirror the causal structure of the world, "judgments about whether two mechanisms are mechanisms of the same kind rely ineliminably on judgments by people (in concert) about the appropriate degree of abstraction required for the problem at hand" (Craver 2014, 589). In other words, depending upon what the aims of inquiry are, be they explanation, prediction or intervention, the way investigators carve up the world will be aligned with their goals.

Even if we acknowledge, as Craver does, that conventional factors play a role in the development of our scientific taxonomies, this does not mean that investigators in neuroscience are not still aiming independently or collectively to discover something like or close enough (from their perspectives) to natural kinds. Craver seems sympathetic to this idea (See for example, Kendler et al. 2011). However, as I aim to show via an investigation of some relevant highlights of the 43-year history of LTP research, sometimes experimental practice is not conducive to the realization of this goal and investigators end up discovering kinds that are closer to the conventional kinds (antirealist) rather than natural kinds (realist) end of the kinds continuum (Craver 2009).

9.3 LTP: One Kind or Many?

Since at least the early twentieth century psychologists and physiologists hypothesized that associative forms of learning required changes in how neurons communicate with each other (See Kandel and Spencer 1968). In 1940, the psychologist Donald O. Hebb, proposed a mechanism for such changes. Hebb claimed that when two cells, A and B, which communicate across a synapse under normal conditions, undergo a period of repeated and concurrent activation, the result is a strengthening of the connection between the two cells, exhibited by a subsequent change in the way cell A excites cell B (Hebb 1949). According to Hebb, each learning event is accompanied by a brief associated activation of two neurons that comprise a synapse, which results in the memory of that event being stored in the form of a physiological change at that synapse.

Although Hebb's proposed mechanism for learning, as one among several "cellular-connection theories of learning" (Kandel and Spencer 1968, 68), had many supporters, during the 1950–1960s, investigators tried to produce activity-dependent Hebbian-like changes at several nervous system synapses (e.g., spinal cord, lateral geniculate nucleus) with limited success.[2] In 1966, however, in the context of investigating the physiology of the dentate gyrus in the hippocampus of adult anesthetized rabbits, Terge Lømo observed an artificially induced physiological equivalent of a strengthening in synaptic efficacy much like the one Hebb had described (See Lømo 2003). After applying a brief yet repetitive stimulation of "one-second bursts of high-frequency (100 Hz "tetanic") stimulation" to perforant path fibers that project from the entorhinal cortex to the dentate gyrus, he recorded, extracellularly, an enduring observable increase in the amplitude above baseline of the evoked field potentials of post-synaptic dentate granule cells. This finding prompted him and his colleagues to further investigate the phenomenon (e.g., Bliss and Lømo 1973; Bliss and Gardner-Medwin 1973) and culminated in the publication of a now famous paper that introduced the scientific community to "a long-lasting potentiation of synaptic transmission" (Bliss and Lømo 1973)—the phenomenon later renamed "long-term potentiation" (LTP).

In order to produce LTP in the dentate area, Bliss and Lømo selected two locations in the hippocampus in which to place stimulating electrodes. One electrode was placed in the lateral perforant path and was used to deliver test pulses (0.1 msec, with maximum amplitude of 100 V) prior to LTP induction. The second electrode was placed in the medial perforant path and used to deliver LTP-inducing stimulation. They selected two locations from which to record extracellularly: (1) the cell body layer, where they recorded population spikes and (2) the dendritic layer, from which they recorded excitatory post-synaptic potentials. Once the stimulating and recording electrodes were at their desired locations in the brain, they applied test pulses consisting of "single shocks at a fixed strength, repeated at intervals of 2–3 sec[onds]" (Bliss and Lømo 1973, 334). Responses were recorded at regular intervals for up to 30 min, "with average responses based on 20 or 30 consecutive single responses" (Bliss and Lømo 1973, 334). A "sequence of conditioning trains" was then delivered at "intervals of 30 min or more". In each experiment they applied

[2] Little work had been done to study cortical synaptic plasticity in the mammalian brain due in part to technological limitations (Kandel and Spencer 1968, 85–86). However, quite a bit of work had been undertaken to induce changes in synaptic efficacy in the invertebrate, *Aplysia depilans* (Kandel and Spencer 1968). Early work on *Aplysia* indicated that activity-dependent changes in synaptic strength primarily involved pre-synaptic (e.g., changes in neurotransmitter release) as opposed to post-synaptic mechanisms (Kandel and Spencer 1968; See also Sweatt 2016).

"one or more conditioning trains" with the trains delivered at "10–20/sec[onds] for 10–15 sec[onds] or at 100/sec[ond] for 3–4 sec[onds]" (Bliss and Lømo 1973, 331).

Bliss and Lømo investigated several parameters of dentate granule cell responses, which they measured, following application of LTP-inducing conditioning trains to perforant path fibers: (1) changes in the amplitude of the population EPSP above average baseline response to test pulses; (2) changes in the peak of the amplitude of the population spike above baseline response to test pulses; and (3) latency of the population spike from post-conditioning test pulses to initial peak of the population spike. Across 15 experiments, they determined that "all three parameters were potentiated in 29% of the experiments" (1973, 331) they conducted. As evidence that they had detected a *bona fide* potentiation, they pointed to the fact that (1) the amplitude of the population EPSP increased in 43% of all experiments, (2) the population spike increased in 40% of all experiments and (3) a reduction in latency of the population spike occurred in 57% of all experiments. Their interpretation of the results was that "two independent mechanisms [were] responsible for [the] long-lasting potentiation" that they observed, namely: "(a) an increase in the efficiency of synaptic transmission at the perforant path synapses" and "(b) an increase in the excitability of the granule cell population" (1973, 332).

Immediately on the heels of Bliss and Lømo's discovery, other investigators began producing LTP in their laboratories. The number of publications directed at understanding the phenomenon and its mechanisms exceeded 3000 by 1999 and 6000 by 2004 (See Sweatt 2016). Before considering some of the details of LTP research during this historical timeframe that is relevant to the question of what kind of kind LTP is, it is relevant to first identify some general features of LTP experiments and relate them to the concepts of natural and conventional kinds described in the previous section.

Experimentation consists of two stages (a) data production and (b) data interpretation (e.g., Woodward 2003). In LTP experiments, data production may be further subdivided into two stages: (1) a design stage in which an experiment is designed and a stimulation paradigm and subprotocol for producing LTP are selected and (2) an implementation stage in which the experimental design is carefully instantiated across each individual experiment. The immediate output of each individual experiment (or implementation of the experimental design) is an individual data point or set of data points. Investigators appeal to these data in order to adjudicate among competing hypotheses about LTP (e.g., hypotheses about its phenomenological features, its synaptic and cellular and molecular mechanisms) (See Sullivan 2009).

As was mentioned above, in the early stages of developing a scientific taxonomy, investigators group together kinds of phenomena that they take to be similar in terms of observable features. Taxonomies are subsequently revised in response to discoveries pertaining to the mechanisms productive of those phenomena. Thinking about the aforementioned features of LTP experiments provides additional insights into the processes that contribute to stabilizing scientific taxonomies and the phenomena to which they correspond. Bliss and Lømo selected an experimental design and sub-protocol for producing LTP. Insofar as across their individual experiments they adhered tightly to this experimental design, they took themselves to be producing, measuring and detecting the same phenomenon across all of the experiments that had identical protocols they undertook in the laboratory. In doing so, they abstracted away from certain differences across these different experiments—different experimental subjects (e.g., rabbits, rats), different times of day in which the experiments were conducted, different stimulating and recording electrodes, different hippocampi, different perforant paths, different granule cells. Insofar as they regarded the data production processes across all of these experiments as reliable, they took themselves to be warranted in grouping the effects produced as instances of the same phenomenon.

Yet, were Bliss and Lømo warranted in grouping the effects produced across their different LTP experiments as instances of the same kind of phenomenon? I think our intuition is to respond "yes" to this question in part because we assume that just so long as the experimental process across these different experiments was reliable, then the mechanisms productive of the effects produced across them were sufficiently identical to be able to warrant classifying them as the same phenomenon (See Feest 2011 on the conditions under which phenomena are stabilized). When scientists make such judgments, they appear to be applying something like the NDR criterion; insofar as they think that the mechanisms across each experiment are identical or conserved, there are no grounds for splitting LTP into as many different kinds as there are LTP experiments. In other words, the scientists regard such abstractions away from the specific details of these experiments as legitimate just so long as they have good grounds for believing that the measures they have taken across experiments do serve to track or preserve real differences. Even though it is clear that conventional factors (i.e., decisions on the part of two investigators as to how to produce, detect and measure the phenomenon) are playing a role in shaping the kinds of phenomena discovered, there seems to be some sense in which the different instances grouped together constitute a *bona fide* kind of phenomenon from the perspective of the investigator(s).

What happens, however, when we consider LTP research taking place after the publication of Bliss and Lømo's 1973 paper? Immediately on the heels of their discovery, other neurophysiologists began producing what they regarded as forms of mammalian LTP in their laboratories. Bliss and Gardner-Medwin, for example, went on to demonstrate that same year that "the same phenomenon" could be produced "without anaesthesia and under normal more stable conditions which can be obtained using chronically prepared animals" (1973, 358). Robert Douglas and Graham Goddard sought to "repeat the observations of Bliss and Gardner-Medwin using the rat instead of the rabbit" but made "several modifications to the procedure" in order to "make the observed potentiation more reliable" (1975, 206). By 1975, other neurophysiologists were producing LTP in their labs, using *in vitro* brain slice preparations (e.g., Deadwyler et al. 1975) and other novel electrophysiological recording techniques, including patch clamp (See Sweatt 2016). By 1976, *in vitro* experiments had been used to produce LTP in three subregions of the hippocampus (CA1, CA3 and dentate gyrus), and "the time course of appearance, magnitude of the effect and duration of LTP appear[ed] to be similar in all 3 areas" (Alger and Teyler 1976, 469). In 1976, Gary Lynch, V. Gribkoff and S. Deadwyler published a letter to the journal *Nature* indicating that "hippocampal synapses" displayed "an unusual degree of physiological plasticity" and that the findings that "modest levels of repeated stimulation cause considerable enhancement in subsequent responses to single pulse stimulation" had "been replicated in several laboratories" (1976, 151).[3]

From the late 1970s to the mid-1980s, the LTP field exploded exponentially. By 1979, LTP at hippocampal synapses had become a model for understanding LTP in the nervous system more broadly and a lot of work was already being directed at uncovering the biochemical processes (e.g., second-messenger signaling cascades) and neurotransmitters (e.g., glutamate) involved in LTP at hippocampal synapses (e.g., Browning et al. 1979; Lynch et al.1979; Dunwiddie and Lynch 1978; Baudry and Lynch 1980; Dolphin et al. 1982, 287). Different investigators used different stimulation paradigms and subprotocols, some that were intended to mimic "naturally occurring firing patterns, observed *in vivo* in the hippocampus" (e.g., theta-burst stimulation) and others that were used to induce LTP "by pairing repeated, single presynaptic stimuli with postsynaptic membrane depolarization" ("so-called 'pairing' LTP") (Sweatt 2016).

I want to assume, for the sake of argument that investigators who engaged in this early LTP research regarded the instances of LTP they were producing in their own laboratories as *bona fide* instances of LTP, much like Bliss and Lomo did, and for similar reasons. That is, with respect to each published research study, just so long as investigators had ensured the reliability of their LTP-producing experiments, they believed themselves justified in thinking that they had stabilized a single kind of phenomenon in their own laboratories. However, if we compare and contrast different features of these experiments—different experimental subjects (rabbits, rats, guinea pigs), different stimulation paradigms and protocols for producing LTP, different preparations (*in vitro*, *in vivo*), different time course with respect to how long the potentiated effect lasted, it becomes difficult to assess whether all of these separate laboratory effects are instances of the same phenomenon or different phenomena. Are the mechanisms across experiments that use different model organisms, different hippocampal synapses or different stimulation paradigms and protocols or different preparations all the same?

In the 1970s and 1980s, investigators did not have answers to these questions.[4] Not enough was known about the cellular and molecular mechanisms of LTP, nor whether cellular and molecular activity differed depending upon the organisms being investigated or stimulus parameters and preparations being used. In response to such uncertainties, some investigators qualified the kind of LTP they were investigating by pointing to differences in the stimulation paradigms, animals and preparations used (*in vivo*, *in vitro*), and/or the synapses at which LTP-inducing

---

[3] The general claim that persisted in the literature until 1986 was that "long-term potentiation [. . .] seen in several hippocampal pathways following repetitive stimulation, [was] somewhat unique when compared to the post-tetanic potentiation seen at the neuromuscular junction or in invertebrates" (Dunwiddie and Lynch 1978, 353–354) in so far that it was longer lasting.

[4] They still lack answers, as I explain later in this section.

stimuli had been delivered.[5] Other investigators, at least prior to 1986, seemed amenable to the idea that despite these kinds of differences, the same phenomenon was under study at least across all hippocampal LTP experiments (e.g., Dunwiddie, Madison and Lynch 1978; Lynch et al. 1979). In other words, there was no real consensus as to how to "lump" or "split" the phenomena. Given that LTP research was, from 1973–1988, in what some reviewers later characterized as "a descriptive phase" (See for e.g., Nicoll, Kauer and Malenka 1988, 97)—a phase in which little was being learned about the mechanisms of LTP—this general lack of consensus about how to "lump" or "split" the phenomena makes sense (See also Craver 2003).

By 1988, LTP research had entered "a mechanistic phase" and results from a series of experiments that involved blockade of N-methyl-D-aspartate (NMDA) receptors at different LTP synapses were taken to establish that LTP at mossy fiber synapses did not require activation of NMDA-receptors (e.g, Harris and Cotman 1986; Staubli 1992).[6] This prompted investigators to split the category of hippocampal LTP into two broad subcategories: NMDA-receptor dependent and NMDA-receptor independent LTP (See also Nicoll et al. 1988; Nicoll and Malenka 1995). Notice that this suggests that there is some sense in which neurobiologists were upholding something like the NDR criterion—given the discovery that LTP could be produced with and without NMDA receptor activation, the hippocampal LTP category was split.

By 1994, investigators had begun to acknowledge that LTP could be induced in many different synaptic pathways by a variety of induction paradigms, and that the biochemical mechanisms of these forms of LTP may differ (See Malenka 1994; See also Lisman et al. 2003). Despite such admissions, some investigators still grouped results from different experiments together in order to make general claims about specific forms of LTP. For example, in 1999, Robert Malenka and Roger Nicoll found it necessary to restrict the focus of their review to LTP induced at "synapses between the Schaffer collateral and commissural axons and the apical dendrites of CA1 pyramidal cells" (1999, 1870). They also acknowledged that because "a review of the literature generates an enormous, even bewildering, list of candidate signal

[5] Bliss and Gardner-Medwin, for example, acknowledged, "the mechanisms of the effect remain uncertain" (Bliss and Gardner-Medwin 1973, 373). Douglas and Goddard were careful to indicate, "this type of potentiation may underlie memory storage in one part of the mammalian brain"5 (1975, 214) rather than at all mammalian cortical synapses. While some investigators regarded "intracellular recordings from *in vitro* preparations of both immature and mature hippocampal tissue" as "similar to those obtained *in vivo*" (Deadwyler et al. 1975, 80), one early failure to obtain LTP in the dentate gyrus of the hippocampus was attributed to the possibility that *in vitro* slice preparations could compromise the integrity of the synaptic pathways (Deadwyler et al., 1975, 84) and result in a "decreased amount of recurring excitation" compared to *in vivo* preparations (Alger and Teyler 1976, 478). By 1978, Dunwiddie and Lynch determined that "various conditioning frequencies apparently induce[d] different degrees of long-term potentiation" (Dunwiddie and Lynch 1978, 366) and that synaptic transmission was required for the initiation of LTP (Dunwiddie, Madison & Lynch 413).

[6] Ursula Staubli, for example, claimed "mossy fiber potentiation is unlike LTP both in induction and expression mechanisms and thus is a wholly different form of synaptic plasticity" (Staubli 1992, 151).

transduction molecules" involved in LTP at these synapses, they were focusing only on those results concerning the mechanisms of this form of LTP that they took to be "compelling" (1999, 1871). By 2003, concerns about whether the mechanisms of different forms of NMDA-receptor dependent LTP (alone) differed depending upon the LTP stimulation paradigms and protocols used were prevalent in the LTP community. Specifically, Robert Malenka pointed to what was then becoming "the increasingly popular hypothesis that different LTP induction protocols result in mechanistically distinct types of NMDA receptor-dependent LTP", which was "often used as a polite explanation for discrepancy in results between [research] groups". Although Malenka suggested that "the evidence in support of this idea remain[ed] weak"; he acknowledged that "it will remain important to seriously consider the possibility that the patterns of activity that are used to elicit LTP influence which intracellular signalling cascades are activated" (Malenka 2003, 925). The following year, Malenka and Mark Bear indicated that "when discussing LTP [...] it is now necessary to define at which specific synapses these phenomena are being studied, at what time point during development, and how they are being triggered (Malenka and Bear 2004, 5). As partial indication that such issues have not yet been fully resolved, Baudry and colleagues recently claimed, "as has been repeatedly mentioned, a major difficulty to integrate all the findings" about LTP "is due in part to the use of different experimental protocols by the majority of research laboratories working on this topic, and the apparent lack of reproducibility of experimental data resulting from these differences" (2015, 74).

Given even these spotty details of the history of LTP, an interesting picture emerges with respect to the kinds of constraints that have to date informed the development of a taxonomy (or lack thereof) of LTP. Over the course of a 43-year history, the published record included well over 6000 papers on LTP. If investigators had begun to treat each independent instance of LTP produced in a given laboratory as an independent phenomenon having an independent mechanism, they would have risked having to develop an unwieldy taxonomy containing as many different kinds of LTP as there were experiments for producing it. That clearly was not practical. So, researchers instead (at least in the context of review papers) abstracted away from specific details of experimental practice that may have led to differences in mechanisms in order to make the goal of providing a unified model of LTP or unified models of specific forms of LTP (e.g. hippocampal NMDA-receptor LTP in area CA1 tractable). Phenomena that may indeed have been produced by different mechanisms thus were sometimes treated as instances of the same phenomenon.

However, notice that by treating different instances of LTP that may have had different mechanisms as all the same phenomenon, researchers abstracted away from potentially relevant causal differences, which seems antithetical to the discovery of kinds that track real divisions in the causal structure of the world.[7]

[7] At the time of Bliss and Lomo's discovery, LTP had been reliably produced in invertebrates across many different laboratories. Bliss and Lomo's finding was different, because it was produced in the mammalian brain. Debates about whether the cellular and mechanisms for LTP induction were conserved across species persisted well into the 21st century (See for example Bickle 2006). It was

That such pragmatic factors were operative in the development of the LTP taxonomy remained hidden until recently, when more and more investigators began to reach consensus that maybe the forms of even hippocampal area CA1 LTP being produced in different laboratories were really different as opposed to the same phenomenon. The reason that they remained hidden, in part, is that there were discoveries, like the discovery of non-NMDA-receptor-dependent LTP that seemed to confirm the idea that LTP researchers were in search of natural kinds and inclined to uphold the NDR criterion. However, as I have demonstrated in this section, a closer look at LTP research reveals that pragmatic factors have played a prominent role in shaping the current taxonomy of LTP.

9.4 Conclusion

At the start of this chapter, I raised four questions to which I want to return in light of the aforementioned analysis. First, do LTP researchers take different instances of LTP produced across different experimental protocols to be the same kind of phenomenon or different phenomena? I think there is widespread recognition that these different instances may not be the same phenomenon and that there is no real consensus about how to "lump" or "split" the phenomena. To cope with the vastness of the experimental record on LTP, some researchers have abstracted away from differences in experimental protocols in an attempt to provide unified mechanistic models of this or that specific form of LTP (e.g., Malenka and Nicoll 1999; Baudry et al. 2015). So, what kind of kinds are LTP researchers are aiming to discover? From one perspective it may indeed be accurate to say that investigators are interested in discovering the mechanisms of LTP and developing a taxonomy that corresponds to real divisions in kinds of LTP. However, the multiplicity of experimental protocols, combined with the fact that discovering mechanisms is a collaborative enterprise are factors that have contributed to neurobiologists having to make pragmatic decisions as to how to lump or split the phenomena.

Acknowledgement The author would like to thank Uljana Feest for helpful comments on an earlier version of this paper and Floh Thiels and Peter Machamer for many interesting and helpful discussions about LTP.

common to hear some investigators arguing that LTP involved pre-synaptic mechanisms and others that it involved post- synaptic mechanisms. Support for conservation of mechanisms waxed and waned depending on the grain of analysis one used to assess similarities and differences in mechanisms across organisms (See Bechtel and Mundale 1999, Sullivan 2009; Craver 2009).

Bibliography

Alger, B., and T. Teyler. 1976. Long-Term and Short-Term Plasticity in the CA1, CA3 and Dentate Regions of the Rat Hippocampal Slice. *Brain Research* 110: 463–480.

Anderson, P., and T. Lømo. 1966. Mode of Activation of Hippocampal Pyramidal Cells by Excitatory Synapses on Dendrites. *Experimental Brain Research* 2: 247–260.

Baudry, M., and G. Lynch. 1980. Hypothesis Regarding the Cellular Mechanisms Responsible for Long-Term Synaptic Plasticity in the Hippocampus. *Experimental Neurology* 68(1): 202–204.

Baudry, M., G. Zhu, Y. Liu, Y. Wang, V. Briz, and X. Bi. 2015. Multiple Cellular Cascades Participate in Long-Term Potentiation and in Hippocampus-Dependent Learning. *Brain Research* 1621: 73–81.

Bechtel, W. 2008. *Mental Mechanisms: Philosophical Perspectives on Cognitive Neuroscience*. New York: Taylor and Francis.

Bechtel, W., and J. Mundale. 1999. Multiple Realizability Revisited: Linking Cognitive and Neural states. *Philosophy of Science* 66(2): 175–207.

Bechtel, W., and R. Richardson. 1993. *Discovering Complexity: Decomposition and Localization as Strategies in Scientific Research*. Princeton: Princeton University Press.

Bennett, M. 2000. The Concept of Long-Term Potentiation of Transmission at Synapses. *Progress in Neurobiology* 60: 109–137.

Bickle, J. 2006. Reducing Mind to Molecular Pathways: Explicating the Reductionism Implicit in Current Cellular and Molecular Neuroscience. *Synthese* 151: 411–434.

Bliss, T., and R. Gardner-Medwin. 1973. Long-Lasting Potentiation of Synaptic Transmission in the Dentate Area of the Unanesthetized Rabbit Following Stimulation of the Perforant Path. *Journal of Physiology* 232: 357–374.

Bliss, T., and T. Lømo. 1973. Long-Lasting Potentiation of Synaptic Transmission in the Dentate Area of the Anaesthetized Rabbit Following Stimulation of the Perforant Path. *Journal of Physiology* 232(2): 331–356.

Bogen, J., and J. Woodward. 1988. Saving the Phenomena. *The Philosophical Review* 97: 303–352.

Browning, M., T. Dunwiddie, W. Bennett, W. Gispen, and G. Lynch. 1979. Synaptic Phosphoproteins: Specific Changes After Repetitive Stimulation of the Hippocampal Slice. *Science* 203(4375): 60–62.

Craver, C. 2002. Dissociable Realization and Kind Splitting. *Philosophy of Science* 71(5): 960–971.

———. 2003. The Making of a Memory Mechanism. *Journal of the History of Biology* 36: 153–195.

———. 2006. When Mechanistic Models Explain. *Synthese* 153: 355–376.

———. 2007. *Explaining the Brain: Mechanisms and the Mosaic Unity of Neuroscience*. Oxford: Oxford University Press.

———. 2009. Mechanisms and Natural Kinds. *Philosophical Psychology* 22(5): 575–594.

———. 2014. The Ontic Account of Scientific Explanation. In *Explanation in the Special Sciences: The Case of Biology and History*, Synthese Library, Vol. 367. Dordrecht: Springer.

Craver, C., and L. Darden. 2001. Discovering Mechanisms in Neurobiology: The Case of Spatial Memory. In *Theory and Method in the Neurosciences*, ed. P.K. Machamer, R. Grush, and P. McLaughlin. Pittsburgh: University of Pittsburgh Press.

Deadwyler, S., F. Dudek, C. Cotman, and G. Lynch. 1975. Intracellular Responses of Rat Dentate Granule Cells *in vitro*: Posttetanic Potentiation to Perforant Path Stimulation. *Brain Research* 88: 80–85.

Dolphin, A., M. Errington, and T. Bliss. 1982. Long-Term Potentiation of the Perforant Path *in vivo* is Associated with Increased Glutamate Release. *Nature* 297: 496–497.

Douglas, R., and G. Goddard. 1975. Long-Term Potentiation of the Perforant Path-Granule Cell Synapse in the Rat Hippocampus. *Brain Research* 86: 205–215.

Dunwiddie, T., and G. Lynch. 1978. Long-Term Potentiation and Depression of Synaptic Responses in the Rat Hippocampus: Localization and Frequency Dependency. *Journal of Physiology* 276: 353–367.

English, J.D., and J.D. Sweatt. 1996. Activation of p42 Mitogen-Activated Protein Kinase in Hippocampal Long-Term Potentiation. *Journal of Biological Chemistry* 271(40): 24329–24332.

———. 1997. A Requirement for the Mitogen-Activated Protein Kinase Cascade in Hippocampal Long-Term Potentiation. *Journal of Biological Chemistry* 272(31): 19103–19106.

Feest, U. 2011. What Exactly is Stabilized When Phenomena are Stabilized? *Synthese* 182(1): 57–71.

Harris, E., and C. Cotman. 1986. Long-Term Potentiation of Guinea Pig Mossy Fiber Responses is not Blocked by N-methyl-D-aspartate Antagonists. *Neuroscience Letters* 70: 132–137.

Hebb, D. 1949. *The organization of behavior; a neuropsychological theory*. New York: Wiley.

Kandel, E., and A. Spencer. 1968. Cellular Neurophysiological Approaches to the Study of Learning. *Physiological Review* 48(1): 65–134.

Kandel, E., and L. Squire. 2008. *Memory: From Mind to Molecules*. 2nd ed. Austin: Roberts & Company Publishers.

Kendler, K., P. Zachar, and C. Craver. 2011. What kinds of Things are Psychiatric Disorders? *Psychological Medicine* 41: 1143–1150.

Lisman, J., J. Lichtman, and J. Sanes. 2003. LTP: Perils and Progress. *Nature Reviews Neuroscience* 4: 926–929.

Lomo, T. 2003. The Discovery of Long-Term Potentiation. *Philosophical Transactions of the Royal Society of London B Biological Sciences* 358: 617–620.

Lynch, G., M. Browning, and W. Bennett. 1979. Biochemical and physiological studies of long-term synaptic plasticity. *Federation Proceedings* 38(7): 2117–2122.

Lynch, G., V. Gribkoff, and S. Deadwyler. 1976. Long Term Potentiation is Accompanied by a Reduction in Dendritic Responsiveness to Glutamic Acid. *Nature* 263: 151–153.

Machamer, P., L. Darden, and C. Craver. 2000. Thinking about Mechanisms. *Philosophy of Science* 67(1): 1–25.

Malenka, R. 2003. The Long-Term Potential of LTP. *Nature Reviews Neuroscience* 4: 923–926.

Malenka, R. 1994. Synaptic Plasticity in the Hippocampus: LTP and LTD. *Cell* 78(4): 535–538.

Malenka, R., and M. Bear. 2004. LTP and LTD: An Embarrassment of Riches. *Neuron* 44: 5–21.

Malenka, R., and R. Nicoll. 1999. Long-Term Potentiation—A Decade of Progress? *Science* 285: 1870–1874.

Nicoll, Roger, Julie Kauer, and Robert Malenka. 1988. The Current Excitement in Long-Term Potentiation. *Neuron* 1: 97–103.

Nicoll, R., and R. Malenka. 1995. Contrasting Properties of Two Forms of Long-Term Potentiation in the Hippocampus. *Nature* 377(6545): 115–118.

Piccinini, Gualtiero, and Carl Craver. 2011. Integrating Psychology and Neuroscience: Functional Analysis as Mechanism Sketches. *Synthese* 183(3): 283–311.

Staubli, U. 1992. A Peculiar Form of Potentiation in Mossy Fiber Synapses. *Epilepsy Research Supplement* 7: 151–157.

Stinson, C. 2016. Mechanisms in Psychology: Ripping Nature at Its Seams. *Synthese* 193(5): 1585–1614.

Sullivan, J. 2009. The Multiplicity of Experimental Protocols: A Challenge to Reductionist and Nonreductionist Models of the Unity of Science. *Synthese* 167: 511–539.

Sweatt, J.D. 2009. *Mechanisms of Memory*. 2nd ed. San Diego: Elsevier.

———. 2016. Neural Plasticity and Behavior – Sixty Years of Conceptual Advances. *Journal of Neurochemistry*. doi:10.1111/jnc.13580.

Watabe, A.M., P.A. Zaki, and T.J. O'Dell. 2000. Coactivation of β-Adrenergic and Cholinergic Receptors Enhances the Induction of Long-Term Potentiation and Synergistically Activates Mitogen-Activated Protein Kinase in the Hippocampal CA1 Region. *The Journal of Neuroscience* 20(16): 5924–5931.

Wimsatt, W. 1981. Robustness, Reliability, and Overdetermination. In *Scientific Inquiry and the Social Sciences*, ed. M.B. Brewer and B.E. Collins. San Francisco: Jossey-Bass Inc.

Woodward, J. 2003. Experimentation, Causal Inference, and Instrumental Realism. In *The Philosophy of Scientific Experimentation*, ed. H. Radder. Pittsburgh: University of Pittsburgh Press.

Chapter 10
Health, Disease, and the Basic Aims of Medicine

Thomas V. Cunningham

Abstract The concepts, *health* and *disease*, have received considerable attention in philosophy of medicine. The first goal of this paper is to demonstrate that three prominent analyses of health and disease can be synthesized if one assumes that medicine is both theoretical and practical, and, therefore, value-laden. The second goal of this paper is to give an account of one route by which evaluative and factual claims come together in medical knowledge, during medical conversations between clinicians and patients. Accomplishing these two goals together yields an epistemology of medicine that moves past debates about health and disease. On this view the conceptual foundations of medicine rest on *subjective* and *objective claims* about health states and their evaluations. By making subjective and objective claims, patients and clinicians negotiate whether patients' health states are sufficiently dysfunctional to warrant medical intervention, leading to the formation of complementary illness narratives, which are anchored to different degrees in personal experience and medical theory. On this view, health and disease are supplanted as the conceptual foundations of medicine. Rather, medicine is better understood in terms of two basic aims: The first aim of medicine is to fully understand patients in scientific and personal terms; the second aim is to intervene upon patients' states of illness in ways that are consistent with full understanding.

Keywords Health • Disease • Facts • Values • Conceptual foundations of medicine

T.V. Cunningham (✉)
Kaiser Permanente West Los Angeles Medical Center, 6041 Cadillac Avenue,
Los Angeles, CA 90034, USA
e-mail: thomas.v.cunningham@kp.org

M.P. Adams et al. (eds.), *Eppur si muove: Doing History and Philosophy of Science with Peter Machamer*, The Western Ontario Series in Philosophy of Science 81, DOI 10.1007/978-3-319-52768-0_10

10.1 Introduction

One of the greatest debates in recent philosophy of medicine concerns the nature of health and disease. Its roots lie in responses to the anti-psychiatry movement of the 1970s (Boorse 1975; yet cf. King 1954), where, guided by the assumption that the concepts *health* and *disease* are basic axioms of medical science (Boorse 1977), scholars aimed to shed light on these axioms in order to clarify the conceptual foundations of medicine. The question of relativism has been central to the debate, where a focal point has been the extent to which assertions about health and disease rest on objective, factual claims about states of affairs or subjective, evaluative claims about what individuals or groups believe are good or bad. This paper aims to repurpose this classic debate over *health* and *disease* by accomplishing two goals.

One goal of the paper is to challenge two distinctions that are commonly used to describe medicine in debates over health and disease. As will be discussed below (Sect. 10.2), philosophers often distinguish medicine-as-theory from medicine-as-practice. Conceptualized as theory, medicine is a science just like any other science. The aim of medicine-as-theory is to generate abstract, generalizable knowledge, rather than knowledge of particular patients whose problems warrant medical intervention. Conceptualized as a practice, medicine aims to identify and categorize patient's illnesses in terms derived from medical theory, yet where the knowledge gained about individual patients is not generalizable and is based on much less evidence than theoretical medical knowledge (Munson 1981). In debates about health and disease, philosophers also distinguish between notions of being "value-free" and "value-laden," drawing on common distinctions in philosophy of science (cf. Douglas 2009). Some claim that medicine-as-theory is value-free, meaning that social norms or values do not influence the knowledge that is accumulated in medical theory. Others hold that medicine-as-theory is value-laden, meaning that social values or norms do influence the knowledge that is accumulated in medical theory. I explore how these two distinctions have been used to describe the conceptual foundations of medicine, as philosophers have argued about the proper scope of the notions *health* and *disease* in theory or practice and their value-ladenness. I then argue that although there are important differences between competing accounts of the conceptual foundations of medicine, received views all share the assumption that if medicine is understood as both theory and practice, then medicine is understood to be value-laden in the sense that social values and norms are inherent to medicine, as theory and practice.

In addition to arguing that if medicine is understood as both theory and practice, then it follows that it must be understood as value-laden, I also propose a richer characterization of what it means for medicine to be value-laden. Drawing on the work of Peter Machamer and coauthors, I show that values and norms can be understood naturalistically as psychosocial entities (Sect. 10.3). Values and norms affect human behaviors, and they may be conceptualized as both *social*, as being held by large groups, and as *personal*, as being held by individual persons. Values and norms may also be stated more or less explicitly: they may be stated explicitly and publicly in language or they may be inferred implicitly from behavior or perceived by a knowing subject. By understanding values and norms this way, it is possible to show how personal values and norms affect knowledge claims in medicine, which is the second goal of the paper. I argue that when patients and physicians discuss patients' health states during medical conversations they are making knowledge claims that include factual claims (about the presence or absence of health states) and evaluative claims (about the value of putative health states). I propose a typology for characterizing these different types of claims—as subjective and objective knowledge claims—and show that characterizing how they are made during the generation of medical knowledge improves our epistemology of medicine because it allows for a better understanding of disputes about medical knowledge while also providing a better way of characterizing the conceptual foundations of medicine than analyses of health and disease. On this account, medicine has two Basic Aims (Sect. 10.4). Aim I concerns the importance of, and roles for, subjective and objective knowledge in medicine. Aim II concerns optimal ways of intervening in patients' health states in light of full knowledge of them.

10.2 Health, Disease, & Values

In philosophy of medicine, there are so-called 'naturalists' and 'normativists' about health and disease. This section describes their views with special attention to how they understand the scope of medicine-as-theory or medicine-as-practice and the role of values and norms in medicine. My goal is to demonstrate that naturalists and normativists agree that if medicine is understood as both theory and practice, then values or norms are necessary for determining whether a person is in a state of disease. In the next section I argue for a richer account of the roles of values and norms in medicine and show how it leads to improvements in our understanding of medical knowledge.

10.2.1 Boorseian Naturalism

The most well known naturalist account has been developed by Christopher Boorse, called the biostatistical theory of disease (BST). As it is so familiar, it can be summarized with a broad brush. On BST, health is defined as the absence of disease, which is defined in terms of reduced organismal functioning relative to other members of a species:

1. The *reference class* is a natural class of organisms of uniform functional design; specifically, an age group of a sex of a species.

2. A *normal function* of a part or process within members of the reference class is a statistically typical contribution by it to their individual survival and reproduction.
3. A *disease* is a type of internal state which is either an impairment of normal functional ability, i.e. a reduction of one or more functional abilities below typical efficiency, or a limitation on functional ability caused by environmental agents.
4. *Health* is the absence of disease. (Boorse 1997, 7–8; cf. Boorse 2013)

Three important assumptions ground BST. The first is that to understand health and disease one should concentrate on medical theory, understood by Boorse as pathology, a composite of many biological disciplines that is distinctly and uniquely focused on disease. The pathologist's conception of disease is concerned with the breakdown of organs and organ systems, discernable in exquisite resolution at the levels of cells and macromolecular structures. Thus, on BST, pathology is the study of differences in functioning of organs and organ systems, which are in turn defined by analyzing organs or organ systems into their component parts and goals and discovering the extent to which some systems and subsystems function more efficiently than others, relative to system goals.[1]

BST's second important assumption is that pathology is value-free. Although Boorse has written extensively about this aspect of the BST, his account nevertheless remains unclear. Boorse claims that BST is value-free, and by this he appears to hold that the concept of disease is value-free insofar as it is defined in terms of the basic science of medicine, pathology. Boorse interprets the property of being value-free in a relative sense, as merely being *as value-free as* certain paradigmatic sciences, so he does not adopt the strongest possible account of what it means to be value-free, where values play no role in the epistemology of a science.[2] For Boorse, the claim that pathology is value-free, means only that pathology is *as value-free as* sciences like "astrophysics" (*ibid.*, 56–58, 75). Boorse interprets this to entail that appealing to human values, desires, preferences, and so forth is neither necessary nor sufficient in order to determine whether something is a disease. On BST, what makes a condition pathological "has nothing to do with how an individual patient views it or how it affects his life activities" (*ibid.*, 46). The patient's views about his health do not matter for pathology, says Boorse; rather, all that matters are the measurements of objective values defined by medical theory. However, this account of

[1] My characterization of Boorse suggests an affinitity with popular accounts of mechanisms within philosophy of science (*e.g.*, Machamer et al. 2000; Bechtel and Richardson 1993). As such, pathology could be described in terms of discovering and explaining biological mechanisms and characteristic types of altered functioning. If so, it might be possible to modify BST account so that functions are defined less in terms of species typicality in the statistical sense that Boorse uses and more in terms of mechanism typicality based on startup conditions, termination conditions, and the entities and activities that bring the former state to the latter, irrespective of the type of organism that is instantiating the mechanism or the typicality with which a mechanism type occurs in species or other evolutionary unit. Developing this insight, lies beyond the scope of the present work (but see Hausman 2012 for a move in this direction).

[2] See Douglas 2009 for a taxonomy of views regarding dimensions of value-leadenness.

the role of values in medicine remains unclear because it leaves open the possibility that if our understanding of other sciences, like astrophysics, is revised to incorporate a robust account of values in science, then this would feed back into accounts of medicine, showing them to be value-laden in the same or similar ways (Cunningham 2015).

A third assumption of BST is that it does not apply directly to medical practice. While Boorse concedes that physician-patient relationships are a basic component of medicine, he argues that those relationships are not a proper component of medicine-as-theory, and thus should not be incorporated into epistemic models of medicine. Rather, by developing BST, Boorse aims explicitly to focus on the allied sciences that make up medical theory, not clinical practices. Boorse argues for separating the concept of "a disease ([e.g.] ulcer or cancer) from the patient's thoughts and fears about it" (*ibid.*, 98), even though he grants that patients' thoughts and fears are relevant to medicine-as-practice. Yet, Boorse claims it is possible to supplement BST and recover medicine-as-theory-and-practice: "BST…provides a theoretical, value-free concept of disease or pathological condition. But on this foundation one can build value-laden disease concepts, by adding evaluative criteria… [Thus,] one can use different 'disease-plus' concepts for different purposes" (*ibid.*, 100). Neither here nor elsewhere does Boorse describe in any detail what a disease-plus-BST would look like. One assumes that it would take into account how physicians and patients interact with one another in clinical practice; since, it is only within the context of actual practice that actual medical knowledge of specific, actual patients is generated.

In sum, BST holds that disease is defined in terms of deviation from species-typical functioning of a part or process of an organism, relative to a demographic population within the species. BST holds that values and norms play no role in how diseases are defined. BST views pathology as the theoretical basis of medicine, which is distinguished from medical practice. And, understanding the epistemology of medical practice is neither necessary nor sufficient for an adequate philosophical account of medicine.

Although BST has many adherents, it appears to be a minority view among philosophers of medicine. The majority view objects to BST's separation of practice—including patients' perceptions of illness and health—from a biomedical perspective of medicine-as-theory. The alternative to BST in the literature is a normativist account of health and disease, of which there are two types, both of which hold that norms and values are foundational concepts in medical epistemology. *Subjective normativists* hold that the concepts of health and disease must be grounded in subjective evaluative experience, where personal values or norms are necessary (but not sufficient) for defining these concepts. *Objective normativists* argue that objective values or norms are necessary for defining health and disease concepts, and they leave open how these objective values relate to subjective evaluative experience.

10.2.2 Subjective Normativism

In its idealized form, subjective normativism is the view that the concepts of health and disease are defined solely in value-laden terms, that is, in terms of individual subjective assessments of value or disvalue. However, scholars rarely, if ever, defend an idealized subjective normativist account; rather, they defend the view that whether a person's symptoms warrant labeling as a "disease" depends upon what that person thinks about the symptoms, and thus, that in order to adequately conceptually analyze the notion, *disease*, requires understanding how individual values or social norms affect the way disease concepts are used in medical practice as well as in medical theory (Cunningham 2014). For example, Rachel Cooper defends a strong subjective normativism when claiming that whether something is a disease depends upon whether the person who is said to be diseased disvalues the health state she or he is experiencing or by the health states entailed by having that disease. Consequently, for her a disease or pathological state in one person may not be a disease or pathological state in another. On her account, a disease is like a weed, in that, "a plant is only a weed if it is not wanted. Thus a daisy can be a weed in one garden but a flower in another, depending on whether or not it is a good thing in a particular garden" (Cooper 2002, 274–275). For Cooper, whether a specific patient is diseased depends upon whether that person disvalues the health states typical of the disease in question.

Cooper's view is not standard, however. When describing their views, subjective normativists distinguish between personal assessments of value and disvalue and social norms, and the extent to which values and norms relate to factual assertions. On Cooper's view, the truth of factual claims about the existence of health states depends upon personal assessments of valuing and disvaluing; and it is not clear what the nature of this dependency relationship is, accept that it arises in medical practice, and, if an individual deems a putatively disvalued state to be neutral (or valued) and consequently rejects the determination that he or she is in a dysfunctional health state, then that person is not properly "diseased." Typical accounts of subjective normativism differ from Coopers view in that they emphasize the extent to which social norms influence the determination of disease, rather than individual assessments of value or disvalue. Typical subjective normativism emphasizes the influence of social norms on how disease entities are characterized in pathology, rather than on how individual values expressed in medical practice influence whether disease claims are endorsed by the patients to whom they refer.

Joseph Margolis provides a paradigmatic account of typical subjective normativism. According to him, health and disease are ascribed using both evaluative and descriptive statements, because descriptive language is logically connected to, and evidentially supported by, vast networks of generally uncontroversial social norms regarding what states of affairs are worthy of rigorous inquiry and explanation. For him, "the concept of disease…makes no sense without reference to appropriate norms" (Margolis 1976, 243), norms that arise from the difficult and complex human activities of describing types of functioning and dysfunctioning in the biological and related sciences. On this view, defining health and disease requires defining functions for parts and sub-parts of human beings. Furthermore, making functional ascriptions requires some appeal to human interests, and is thus at least partly evaluative because of the influence of social norms on scientific categories, methods, funding practices, and so on. On Margolis' account, although concepts in pathology may be universally accepted as describing dysfunctional states (e.g., cardiac arrest, pulmonary embolism), for other functional ascriptions there may be little consensus on what it means to be dysfunctional. For example, definitions of hypoactive sexual desire disorder (Montgomery 2008) will be controversial because defining thresholds for "appropriate" sexual desire will be controversial because of the lack of consensus on social norms regarding sexuality. In such cases, the evaluative aspect of medicine will be more salient because social norms will determine which behaviors are deemed functionality adequate and which are deemed deficient. Thus, according to Margolis, "medicine is a doubly normative discipline" (*ibid.*, 242), because social norms for scientific inquiry influence which types of dysfunction are investigated scientifically and therefore become well understood; and, social norms regarding what it means to be dysfunctional influence which types of behaviors are construed as dysfunctional when they are investigated scientifically.

On both the radical and typical versions of subjective normativism, values and norms play a necessary role in constraining how patients' health states are determined to be sufficiently dysfunctional to warrant labeling as being in a state of disease. On radical subjective normativism, individual values constrain the determination of disease in the context of negotiations between physicians and patients in medical practice. On typical subjective normativism, social norms constrain the determination of disease because they influence the construction of disease categories in medical theory.

10.2.3 Objective Normativism

Objective normativism also emphasizes the importance of values and norms in determining whether someone is diseased. Yet, unlike subjective normativism, it holds that medicine is value-laden because it appeals to values and norms that are objective in some way. For instance, James Lennox argues that BST only appears to be value-free because it "reduces the apparently evaluative basis of medical judgments to statistical judgments" (Lennox 1995, 500). He argues instead that there is at least one basic, objective value for all living things, namely, life. Thus, attribution of health is both objective insofar as it is the attribution of a function's contribution to continued unencumbered life, and it is value-laden, insofar as it rests upon a prior commitment to a basic value of existence. Consider the case of the relationship between elevated cholesterol levels and cardiac arrest. According to Lennox, we are interested in cholesterol levels not because they are typical of the species, as BST stipulates, but because they are predictive of malfunction, of arrest, of death.

Knowing how the circulatory system contributes to a person's persistence is integral for understanding why measuring cholesterol levels is important for health. Regardless of whether a specific range of measurements is statistically normal, certain values are predictive of poor functioning. Thus, by emphasizing that the biological notion, *life*, is value-laden, Lennox demonstrates that BST errs when presuming that claims about pathophysiology are completely value-free. Lennox's account emphasizes the role of objective values in how dysfunction is defined when medical theory categorizes disease processes and entities.

Yet, Lennox's view is open to the same line of criticism that subjective normativists lodge at BST, that by privileging one value as the basis of all functional ascriptions it unjustifiably reduces the basis of value, and thus, how states of health and disease are ascribed. Rather than emphasizing species typicality as the basis of all functional ascriptions, so this response goes, it emphasizes contribution to existence, which is ultimately based upon ascriptions of differential fitness. But, like other organisms, humans have various interests and goals; and thus, although from an evolutionary biological point of view the only goal that matters is relative increase in fertility, from a broader point of view, there are other human values and norms of importance on a par with life and reproduction (Ereshefsky 2009).

Ken Schaffner provides a response to this objection. He suggests that a broad suite of human values may be objectively described, and thus, that Boorse's account may be extended in a way that makes it inclusive of normative content while also remaining objective in a sense that is consistent with other features of the BST. Schaffner claims human sociality provides a foundation for a sentimentalist theory of value that makes the multiplicity of human values intelligible in objective terms. Starting from game-theoretic models of human evolution, he argues that evolutionary forces are likely to have resulted in selection for individual- and community-level values like cooperative behavior (cf. Sterelny 2016). Invoking Hume, such behavior, he says, "will be grounded in dispositions to behave cooperatively with other humans, that *from a subjective, motivational point of view*, will be seen as *sentiments*, including those of sympathy, empathy, and benevolence" (1999, 370). Thus, according to Schaffner, a collection of dispositions to feel sentiments like sympathy and empathy, and to behave in ways that follow from them may have been selected for; and so, a moral sense may too have been selected for. But looked at from the standpoint of selection, these values may be explained by appeal to evolutionary theory and fitness, just as other traits can be explained. Consequently, normal functional abilities, so defined, include sentiments: "These sentiments are also moral or evaluative notions, and thus can be the basis of a prescriptive or valuational element in a now modified and extended Boorsean account" (*ibid.*, 371). Schaffner's view thus echoes a both typical and radical subjective normativism. In his account, a basic suite of individual values has been selected for and is thus objectively explicable, like other evolved aspects of human behavior. Schaffner also holds that social norms may also be selected for,[3] so his view is consistent with a

[3] Group/individual selection.

typical subjective normativism, yet, with the additional claim that such values may be objectively explicable via scientific inquiry.

On Schaffner's extension of BST one finds a route to recovering values within BST. One may adopt an objective account of medicine-as-pathology and yet still examine pathological social and moral behavior at the levels of persons, community groups, regions, societies, and so forth (cf. Engel 1977). At a certain point, these "dysfunctions" may be so well described scientifically that they are consistent with a BST account of disease; hence, one may be diseased in ways that are perfectly suited to moral discourse, to description in emotive, social, ethical, and moral terms. If so, then it is not at all inconsistent to place moral considerations *within* the biostatistical theory.

10.2.4 Theory and Practice, Facts and Values

Surveying naturalism and normativism about health and disease reveals disagreements about the proper scope of medicine and the appropriate way to understand the roles of values and norms in medicine. Naturalism defines medicine narrowly in terms of medical theory and pathology, and holds that values fall outside of a narrowly construed account of medicine-as-theory and into a vaguely defined region of medical practice. Radical subjective normativism holds that the claims of individual persons about values are a part of medicine, as they arise in the context of medical practice. Typical subjective normativism holds that social norms are a part of medicine, as they constrain which disease entities are sufficiently understood scientifically to be included in medical theory and which states are deemed dysfunctional by society. The most promising account of objective normativism holds that both claims of individual persons about values and communal or social norms are part of medicine, because they influence what health states are viewed as dysfunctional, both in the context of medical practice and the entities and processes characterized by medical theory.

Abstracting away from these disagreements, it is possible to see an overarching agreement among all of these views: naturalists and normativist both agree that *if medicine is understood as a composite of theory and practice (rather than merely as medical theory), then medicine is value-laden.* Recall that Boorse offers a "disease-plus" sketch of the BST, where he presumes that a defensible model of the way values and norms impact medical practice could be fruitfully grafted onto BST. Likewise, taking both radical and typical subjective normativism together as one composite view entails the view that personal values and social norms both influence the way that disease entities are determined, either at the level of individual claims about disease states or at the level of how diseases are scientifically categorized. As just described, Schaffner's objective normativism is explicit about the influence of individual values and social norms on medicine. Though he is not explicit about the relationship of the theory/practice distinction to his account, since his view emphasizes individual claim-making as well as theory-building, it is implied that he views medicine to encompass both medical practice and medical theory.

Having uncovered this shared agreement by naturalism and normativism about the importance of values and norms to medicine if the scope of medicine is broad, the rest of this paper aims to describe the meaning of values and norms in additional detail, and then to show that this view yields an alternative epistemology of medicine that integrates the emphasis on the personal level of resolution central to radical subjective normativism and objective normativism and the emphasis on a trans-personal level of resolution found in typical subjective normativism, objective normativism, and naturalism about health and disease. In addition to integrating these different emphases, I argue for supplanting the concepts, *health* and *disease*, as the proper conceptual foundations of medicine. On my view, subjective evaluative assessments and objective ascriptions of function and dysfunction are both integral to medicine. These assessments and ascriptions come together in the process of assessing patients' health states and proposing medically appropriate interventions to change them. The conceptual foundations of medicine can be understood in terms of two basic aims, of understanding patients and intervening to alter their health states, which require adequate subjective and objective understanding.

10.3 Knowledge Claims in Medicine

To clarify the roles that values play in medicine, hereafter understood as both theory and practice, I begin by defining values and norms. Then I consider how they are asserted during medical conversations between clinicians and patients about health care. I then claim that thinking about how values and norms arise during conversations where factual claims are also asserted leads to a novel schema for understanding the conceptual foundations of medicine. On this view, medicine may be understood in terms of two types of knowledge claims, subjective claims expressing evaluations and objective claims expressing facts. Subjective claims express individual persons' evaluations of putative health states. Objective claims come in degrees of objectivity, and express agreed upon characterizations of health states and their value or disvalue, which is constrained by group norms.

10.3.1 Machamer on Values and Norms

To define values and norms, I will appeal to work by Peter Machamer and coauthors. In a paper with Douglas, Machamer writes that values "are reflections of what people find important or valuable" (1998, 31). In addition, he says that values may be understood psychologically, in relation to judgment, action, motivation, and perception. For Machamer, values play roles in how individuals and groups determine which actions to perform out of available options: "People evaluate a problem or course of action by generating a set of possible actions and drawing their implications. They then select an alternative from this set of possibilities on the basis of the value of the action and its likely consequences" (*ibid*). On this view, values are expressions that summarize possible actions, coupled with potential outcomes. Values are summary in that they reduce a complex set of actions and outcomes to simple expressions of positivity (good) or negativity (bad).

Values and norms are closely related. Whereas values are summary expressions of what individuals, groups, or institutions find important and valued or unimportant and disvalued, norms are guides for how to achieve valued ends and avoid disvalued ones. Norms are directed at outcomes or methods of achieving outcomes; they serve to express ways of achieving what is valuable or of avoiding what is not. Norms may be explicit or implicit. Again, following Machamer:

> *Norms* are like rules or the principles that guide regulated activities. They are embodied in performance skills; they are constitutive of knowing how to do something, be it intellectual, practical, or productive. They apply to individuals or groups. They apply in specific social locations. And they are public: they can be learned and observed and have the possibility of being used to correct actions or practices that do not accord with them. (Machamer and Wolters 2004, 9).

On this view, norms are necessary for knowledge, as they provide guidance for how to perform acts requiring skill. Norms may be shared by groups of persons or adopted by individuals. Most important, norms may be public: one can observe norms by attending to what others say and do, one can be taught norms, and one can act in ways that are more or less consistent with norms.

On this account, values and norms may be private or public. Privately, they may be understood as cognitive motivational states that may be articulated, assessed, and evaluated internally by a subject. Publicly, values and norms are expressed in human behaviors that imply them or explicit statements of them to others in language.

10.3.2 Values and Norms in Medical Conversations

Machamer's view of values and norms is useful because it describes values and norms as psychosocial entities. They are features of the world that are constitutive of how persons think, communicate, regulate behaviors, and influence one another. And they can be observed in action or by explicit statement to others in language. Since the account defines values and norms in terms of their functions in human cognition, behavior, and language, it provides a way of talking about values and norms that clarifies the relationship between assertions of values at the personal level of resolution and norms functioning at the trans-personal level that influence how individuals conceptualize, and talk about, their experiences. Below, I build on Machamer's approach by describing how claims about the value or disvalue of health states arise in medical conversations. This account of how values are stated, disputed, and eventually accepted or dismissed captures a role values play in influencing how medical knowledge is made. In addition, by describing how norms influence the acceptance or rejection of evaluative claims about putative health care states, I also show how norms influence medical knowledge about functional and dysfunctional states.

On Machamer's account, values and norms may be stated explicitly to other individuals or perceived personally. Whether state explicitly or perceived, values and norms may function as knowledge claims. In medicine, this is important because it is possible to distinguish between two different routes by which values and norms generate knowledge claims, and thus, two different types of knowledge claims.

When an individual is in a state of valuing, disvaluing, or recognizing a norm, but does not state this perception to another person, then this constitutes making a private, implicit, claim. This is one way knowledge claims about values and norms occur, which I propose to call *subjective claims* about values and norms. Knowledge claims about values and norms also occur when they are asserted to other persons, in which case they are *objective claims* about values and norms because they—by virtue of being made public—then become open to scrutiny. As claims are explicitly stated to others, they become corrigible claims to knowledge that may be further objectified in terms of their consistency with other claims, plausibility under available evidence, predictive accuracy, precision, and so on.[4]

To see the difference between subjective and objective claims in medicine, imagine, for example, that one perceives he has been drinking too much wine and it is harming him. Such an assessment is both factual, in that it posits a fact of the matter about the individual's drinking, and evaluative, in that it stipulates an assessment of that behavior as harmful and bad. Imagine further that one does not share this view with others. One may momentarily note this or obsessively reflect on it for years. Regardless of the time scale over which this claim is held by the subject, this would be a paradigmatic instance of a subjective claim about one's health: it is a claim about a fact or set of facts, coupled with an evaluation or set of evaluations of the fact or facts, and it is not shared with others who can scrutinize it.

Imagine further that, in this example, one shares a private judgment with a clinician. When one shares this concern publicly, giving language to it, then the evaluation undergoes a process of objectification. Those who then have knowledge of the judgment have an opportunity to concur with the evaluation, to argue for its revision, or to outright reject it. Imagine one says to a primary care physician that he "drinks a lot" and believes "it is making me sick." The physician might ask for clarification, wanting to know just how large those drinks are, what they contain, how frequently they are consumed, and what symptoms are felt by the speaker. In this way, the physician may come to concur with the judgment that the reported behavior is excessive and dysfunctional, rising to a threshold that warrants deploying a clinical diagnostic term like "alcohol use disorder" (APA 2013). Or, the physician

[4] I use the concept of corrigibility following Suppes' account, where he distinguishes between corrigible and incorrigible data in scientific experiments. On this view, subjective claims are incorrigible because they are accepted without challenge and are not open to scrutiny, save for personal, introspective skepticism. A hallmark of objective claims is that they are open to challenge and scrutiny, just like corrigible data (Suppes 1977, 277 *ff.*).

may come to believe that the behavior does not rise to a clinically meaningful level warranting categorization as a diagnosable dysfunction. In making this discrimination during conversation with a patient, the physician scrutinizes the patient's claim, probing it for consistency with assumptions about the theoretical construct, "alcohol use disorder," and consistency with other claims the patient makes or other knowledge the physician has about the patient's medical history. In this way, as a subjective claim is made public, the claim may also become an objective claim, nested within complex claims about disease states, symptoms, lay-speak and scientific language, pathologies, statistics, and so on. Thus, there is a route between the subjective assessments Cooper argued were central to making a disease determination and objective claims made by persons being evaluated for, or receiving, medical care. In this fashion, subjective and objective claims can not only co-occur, but also often do, because there are crucial moments in medical practice where making objective claims about a patient requires soliciting information from him about his assessment of a deviation from health.[5]

10.3.3 Synthesizing Naturalism and Normativism About Health and Disease

Thinking of evaluative claims in terms of two distinguishable types—as either subjective or objective—provides a way of synthesizing normativism and naturalism about health and disease. When clinicians and patients talk about health care, clinicians solicit patients' evaluations of their experiences: they ask patients what is happening to them, whether (and why) it is bad, and what their baseline experiences are. Such conversations convey what, for a patient, constitutes a sufficiently significant deviation from normal experience to warrant presenting to a physician.[6] If such conversations are successful, then patients describe what is valuable in their own terms: they make private, subjective claims about their experiences public, and thus make objective claims about their health to clinicians. During medical

[5] In his 1984, *The Discourse of Medicine: Dialectics of Medical Interviews*, Mishler analyzes this activity of objectification in a typical process of taking a medical history. Using a similar example involving alcohol use, Mishler details how, during a typical physician-patient conversation, the patient's "life world" is partially assimilated with a "world of medicine" as the physician constructs a quasi-scientific rendering of the patient's experience in language consistent with medical theory and practice. At the same time, Mishler also shows how the "world of medicine" is fed back into the patient's way of speaking of her experiences, and thus, during objectification the patient takes up ways of speaking that originate in medical theory and practice, making these locutions her own.

[6] This ignores cases where a patient cannot speak for him or herself due to incapacity. Although physicians have a duty to incorporate patients' values and preferences into decision-making when they are incapacitated (Buchanan and Brock 1989; Scheunemann et al. 2012), empirical research suggests they often do not (Scheunemann et al. 2015). For sake of simplicity such cases will lie outside of the present discussion.

conversations, clinicians also aim to reconstruct patients' experiences in terms that are consistent with medical theory and create opportunities to empirically evaluate them. This reconstruction of the patient's circumstances may be more or less consistent with the patient's own stated knowledge of her condition. It will also typically differ from the objective claims a patient makes because it will aim for precision in measurement in order to fit with medical theory. Thus, where a patient may refer to "drinking a lot," a clinician may clarify the exact number and types of drinks per day, as captured by Mishler in his classic (1984) analysis of medical conversations. This information may then be entered into the medical chart in a shorthand that increases its objectification through precision and use of jargon; thus, "drinks a lot" becomes "drinks: 4–5/d" in the chart (cf. Berg 1996). In this fashion, patients' objective claims are further objectified as clinicians and patients settle on language that captures aspects of the patient's experience.

Through this process, theory and practice intersect in medical conversations, which produces parallel and complementary narratives about the patient's circumstances. While patients' objective claims remain anchored to their subjective experiences and private subjective claims, physicians' objective claims balance verisimilitude with characterizing the patient in theoretical terms. Thus, neither naturalism nor normativism adequately characterizes how knowledge is made in medicine. Rather, each view partially captures the generation and flow of medical knowledge, depending upon its emphasis on subjective versus objective assessments of dysfunctional or functional states, and the attending value or disvalue of those states. But by focusing on how subjective and objective claims are made—and how they function together during the process of representing patients' health—it is possible to put these differential emphases together into a single, overarching account of medical epistemology. The foundation of medical knowledge rests in medical conversations, the point at which persons' private, subjective assessments of dysfunction or function are made public, at which a process of objectification of those assessments begins. Medicine is about health and disease; however, subjective claims have a certain priority over claims about health and disease, and subjective claims are most well understood in terms of their relationship to objective claims, rather than to claims about health and disease.

This way of thinking about medicine is indebted to Mark Ereshefsky's response to debates about the concepts of health and disease. Ereshefsky argues against both naturalists and normativists, holding instead that a better way to think about health and disease is to think in terms of state descriptions and normative claims, where state descriptions are "descriptions of physiological or psychological states" and normative claims are "claims about what states we value or disvalue" (Ereshefsky 2009, 221). Like me, Ereshefsky aims to reorient discussions about health and disease toward the issue of *how* normative and state-description claims co-exist when medical knowledge is asserted, rather than *whether* they co-exist.

Using deafness as an example, Ereshefsky makes the point that some contend deafness is a disease, and by doing so, contend that it follows that all deaf persons warrant treatment, such as with cochlea implants. Others, however, argue deafness is not a disease, holding that deafness confers advantages, thus marking a beneficial physiologic difference rather than a 'disease.' Ereshefsky's response is to point out that focusing on normative and state-description claims sheds light on the issues whereas focusing on definitions of health and disease does not. This is because, as he says, "framing the debate [in terms of health and disease] masks points of agreement and disagreement between the two sides. Both parties agree that there is a physiological state involving hearing, but they disagree over whether such a state should be valued or disvalued" (*ibid.*, 225). On my view, Ereshefsky's account can also be understood in terms of knowledge claims. So understood, his view is that using a typology of normative knowledge claims and state-descriptive knowledge claims—rather than the attributes of being healthy versus being diseased—illuminates where the disagreements lie while identifying areas to work toward compromise or resolution.

While Ereshefsky's view is an important move forward in the epistemology of medicine, I have offered additional important details to the story. Medical knowledge claims may be understood in terms of whether they describe states or evaluations, as well as in terms of their objectivity and subjectivity. Doing so improves understanding by increasing the resolution at which disagreements and agreements of substance can be identified, because it allows one to describe other moments where disagreements arise, such as, the moments where clinicians evaluate patients, listen as patients make objective claims about their experiences, formulate objective claims in response to patients' claims, and scrutinize those claims in processes of objectification. In such moments, because of their authority and expertise, clinicians describe patients in ways that the patients may not understand due to low medical literacy or that patients may disagree with precisely because they reject the implications of being so categorized (e.g., the case of deafness). At such times, clinicians also describe patients by making objective claims that will later be scrutinized and evaluated by other clinicians for their accuracy, consistency, plausibility, evidential warrant, scope, and so on. But, in creating this alternative description it is not the case that patients' descriptions of their experiences thereby disappear. Patients' subjective conceptions of their illness experiences remain, persist, and fluctuate. Kleinman, for example, has termed these "illness narratives" and argues they are central to understanding the patient's experience of illness and its appropriate role in medical knowledge (Kleinman 1988, cf. Frank 1995).

The relationship between patients' and clinicians' claims about patients' health states and need for care is nicely captured by distinguishing objective claims from subjective ones. A distinction between subjective and objective claims also combines well with Ereshefsky's account to provide a richer conception of medical knowledge than the normativist or naturalist accounts do while remaining consistent with those frameworks. Coupling Ereshefsky's view with the one argued for here yields a typology of medical knowledge in terms of four types of claims. Claims about the *origins and corrigibility of knowledge* may be categorized in terms of subjective or objective claims. Claims about the *purpose of knowledge* may be categorized in terms of state descriptions or normative claims. The resulting account fruitfully moves past the issues of what health and disease are and whether either or both are axiomatic because it can be drawn on to provide an alternative conceptual analysis of the foundations of medical epistemology than those concepts. This is the aim of the penultimate section.

10.4 The Basic Aims of Medicine

The vision described here is one of a multifaceted epistemology of medicine. I describe the diversity of ways that patients and clinicians make, justify, contest, and accept knowledge claims in health care. On this account, medical knowledge is gained as patients and clinicians formulate and perceive private, subjective claims about health states and their value or disvalue, and when patients and clinicians articulate these claims publicly, in the form of objective claims about patients' health states and normative claims about their value or disvalue. On this view, the basic components of medical knowledge are the perceptions of patients and clinicians and the language they use to describe them in processes of objectification. During these processes, incorrigible, private observations are made public and, therefore, become corrigible claims that may be disputed and agreed upon. The foundations of medical knowledge are thus to be found in these processes, rather than the concepts, *health* and *disease*. In this penultimate section, I argue that there are two basic aims of medicine, which, through their complexity, capture the ways that medicine aims to understand patients' circumstances and intervene upon them appropriately.

From an epistemic perspective, medicine may be understood as having two basic aims. One aim of medicine is to understand the patient objectively and subjectively; the other is to intervene by treating the patient via means that are consistent with his state of health, defined in terms of the understanding gained if the first aim is met. Schematically, the two Basic Aims of Medicine are:

Aim I: To fully describe the patient's circumstances objectively (in scientific terms) and subjectively (in personal terms).
Aim II: To treat the patient via means that are consistent with the patient's health state defined in Aim I.

Describing the foundation of medical knowledge in terms of two basic aims captures the insight shared by participants in debates over health and disease that a fundamental goal of medical theory is to scientifically describe patients' conditions, or, as it is understood on the biostatistical theory, to describe their dysfunction in terms of deviations from population-relative levels of normal functioning. Aim I is also consistent with the insight that a fundamental goal of medical practice is to adequately describe patients' perceptions of health, even to the extent that their perceptions are inconsistent with medical theory. Together, the Basic Aims also clarify the interventional goals of medicine and the path-dependency of medical knowledge. In addition to the fundamental goal of gaining full knowledge about patients' conditions (Aim I), medicine also aims at intervention to change patients' health states, once they are fully understood. Thus, Aim II makes it explicit that a basic goal of medicine is to modify patients' health states. Unlike conceptual analyses of health and disease, which focus on the ways that pathological conditions are determined and the extent to which values and norms are relevant to their determination, Aim II broadens the scope of medical knowledge explicitly to include modifying patients' states of being. Thus, medicine aims not only at fully depicting the extent to which patients are in states of disease, but also at acting to change those states, when justified by factual and evaluative knowledge.

This way of characterizing the fundamental goals of medicine follows if, as has been argued above, our conceptual analysis of medicine is directed away from whether *health* and *disease* are adequate conceptual foundations and instead focused on how subjective and objective claims are made when describing patients' health states and evaluating them. In addition to the benefit that the account repurposes debates about health and disease toward *how* claims about facts and norms come together in medical theory and practice, this account of medical knowledge is also beneficial because it draws on longstanding issues in philosophy of science and recent progress that has been made on them in order to make progress on fundamental conceptual issues in philosophy of medicine.

Historians and philosophers of science have long discussed the roles of objectivity and subjectivity in science. Consider, for example, Quine's classic account of how evidential claims are made in the context of scientific inquiry. Quine describes scientists who assert what he calls "observation sentences," which may be understood as claims that garner assent from fellow observers partaking in inquiry. Importantly, observation sentences serve "as vehicles of evidence for our knowledge of the external world" precisely because they motivate agreement between persons, and despite the fact that assenting to them "may be more or less delayed or hesitant" and may be corrigible, or open to "self-correction" (Quine 1993, 108–109). According to Quine, such sentences provide an ideal basis for scientific inquiry because they relate both to what he calls "the subjective or solipsistic half" of the story and the "intersubjective" half: "The observation sentence is Janus-faced. It faces outward to the corroborating witness and inward to the speaker. It faces outward to its subject matter and inward to the range of neural intake that is keyed to trigger it" (*ibid.*, 109).

Most importantly, observation sentences garner assent from persons who engage in shared perceptions of states of affairs. Consequently, they are just as useful for describing assertions in the context of medical inquiry as in scientific inquiry. In medicine, there are always at least two persons involved in inquiring about a patient's health, the patient and the clinician. Each observes the phenomenon of deviation from health from a different vantage point, in that patients rarely view their experiences in terms of medical theory and clinicians cannot view a patient's experiences from a subjective point of view. When physicians and patients make claims about health states and their value or disvalue during medical conversations, the claims may be understood as observation sentences made in the context of inquiry about health states and their values.

For the patient, observation sentences are about perceptions of deviation away from a baseline state of health. When making claims about their experiences, patients use language that need not be consistent with medical theory. They make claims about health in their own, *personal terms*, in language that is meaningful for them. As patients convey their perceptions of illness, physicians also form conceptions of the patient's experience in terms that transcend medical theory. Thus, patients and physicians both strive to conceptualize the patient's circumstances in personal terms. The notion, "personal terms," is intended to capture psychosocial behaviors that arise from social interactions, including the feelings of empathy, sympathy, and benevolence remarked on by Schaffner (1999). Yet merely *feeling* empathy for a patient's condition does not suffice, which is why I use the locution "personal terms" rather than merely the adverbial phrase, "understanding personally." That is, on this view, understanding a patient in personal *terms* happens if and only if there is an attempt to describe that patient's experience in language given by the patient.

Observation sentences also capture what it means to describe a patient in scientific terms. In addition to conceptualizing patients' experiences in terms that are familiar to them, clinicians also aim to describe patients' experiences by making observations during clinical examinations and by collecting many lines of evidence supporting or countering those examinations, including patients' own reports of their experiences in medical conversations. When describing patients' experiences in medical terms, clinicians aim to capture the patient's health state using language that is consistent with medical theory, using terms designed for corroboration, terms that invite agreement or disagreement, and in doing so, contestation and inquiry as to their appropriateness. To understand patients in medicine requires *objective understanding*, or understanding in scientific terms. Objective understanding occurs as individual knowers observe states of affairs and make claims about them. These claims are based on observations—and are thus observation sentences—yet they are also, by virtue of their conceptualization by a knowing subject, formulated in language that is designed to be consistent with medical theory. In this way, objective claims in medicine function to relate the observed behavior of specific patients to medical theory about patients in general.

As philosophers of science have recently argued, scientific knowledge is interventionist, in that it aims to depict systems in ways that reveal the causal relationships that produce changes within them, and thus, ways that manipulating the system will predictably change it (Woodward 2003). Like scientific inquiry, medical knowledge aims at intervention. Medical knowledge is gained for a specific purpose, to facilitate appropriate manipulation of patients' bodies and behaviors in order to alter their states of health. Basic Aim II captures this insight about medical knowledge well. Yet, it does so in a way that anchors intervention to *full* understanding, as captured in the first aim. It captures the intuition that intervention is always a means to an end, and that end is to help a patient without doing harm. Here, harm is captured in both subjective and objective terms. Harm would be done if an intervention is performed that is inconsistent with how the patient is understood scientifically. If the patient was experiencing an asymptomatic inguinal hernia, for example, then surgical repair would be inconsistent with a scientific understanding of the patient because it has a much higher risk of increased pain than the dysfunction itself and because it also has attending low risks of far more severe outcomes than closely watching the dysfunction over time (Simons et al. 2009). Likewise, harm would be done if an intervention were performed that is inconsistent with a patient's subjective claims about his or her state of health. For example, take the fictional case of a motorcyclist who presents in the emergency department after a devastating accident (Aulisio 2003). Her right leg is in danger of requiring amputation due to decreased blood flow, so the surgeon recommends moving a vein from her wrist to her leg, making amputation unnecessary. Going forward with surgery seems patently rational, however, unbeknownst to the surgeon, the patient is also a concert pianist whose self-identity and livelihood depend on being a musician. Consequently, her personal assessment is that losing the leg is preferable to altering her wrist. As such, even if the operation were successfully performed, it would harm her because it resulted in circumstances that were markedly inconsistent with an adequate subjective understanding of her experience of illness and predictable consequences of different medically appropriate interventions.

10.5 A Naturalized Epistemology of Medicine

In their analyses of health and disease, most philosophers of medicine have assumed that there is something so special about subjective assessments of health that such assessments must stand outside the scope of medical knowledge. Above, I have offered a competing analysis, which aims to show that by focusing on medicine as practice and theory, rather than medicine-as-theory, it is possible to synthesize naturalism and normativism, and, moreover, that this synthesis yields a novel epistemology of medicine.

On my view, medicine is simultaneously both theoretical and practical. Thinking of medicine in this way shifts attention to how values, norms, and factual assertions come together during the formulation of medical knowledge, rather than whether factual and evaluative notions can be cordoned off into separate logical realms, and then whether either of these realms should properly be identified as part of medicine. I have argued that distinguishing between objective and subjective claims about states of health and normative evaluations of them provides a means of characterizing medicine, understood as encompassing both theory and practice. Furthermore, I have also shown that with these distinctions it is possible to articulate alternative conceptual foundations of medicine. Rather than looking to the notions, *health* and *disease*, as axioms that define the epistemology of medicine, I have argued that medicine has two basic aims, to fully understand patient's circumstances and to intervene upon them in ways that are consistent with full understanding.

A strength of my position is that it provides a new option for characterizing the role of values and norms in the conceptual foundations of medicine. My account differs from a paradigmatic subjective normativism, where values and norms are understood to function as upstream constraints on the types of objective claims that can be made about disease. On the account given here, evaluative claims are constitutive of medical knowledge because they arise during medical conversations and, thereby, constitute knowledge that clinicians must account for when conceptualizing a patient's health or dysfunction. My view also differs from a radical subjective normativism, where, for a health state to qualify as a disease, it is necessary that the patient evaluate it as bad. The account given here concurs with the emphasis on patients' subjective claims about their own health states but without stipulating that patients' agreement with clinicians' objective claims about dysfunction is either necessary or sufficient in the determination of disease. Instead, the account proposed here defines subjective knowledge claims and theorizes how they lead to objective knowledge claims about health states and evaluations of them. Basic Aim I of medicine requires *considering* patients' subjective perspectives to the extent possible. If patients are unable to express their perceptions of illness, or if patients' perceptions deviate in specific, predictable ways from objective measures of illness, then on this view a patient may still be described as dysfunctional. Likewise, in cases where a patient's perceptions diverge from the observations of his clinicians, the patient's perceptions may trump, and thus the patient may not be viewed as dysfunctional, or may be viewed as dysfunctional but in ways that fail to justify intervention.

There are also limitations to my view. In stating my account, I have argued for a naturalized epistemology of medicine that is most consistent with an objective normativist account. Like Schaffner, I believe that the biostatistical theory of disease is a good explication of what it means to be in a dysfunctional health state, but that the BST fails to adequately accommodate the importance of human sentiments, values, and norms in medicine. To capture how medicine accommodates these concepts, I have emphasized regularities I observe in medical practice, where, in conversations and examinations, physicians and patients construct alternative narratives to explicate the patients' circumstances. In this way, my view is more fine grained than Schaffner's, because it rests on assumptions about empirically observable regularities in the clinical practice setting, whereas Schaffner's account rests on assumptions about evolved dispositions or capacities shared by all human populations. Both Schaffner's and my accounts suffer from their vagueness, in that they articulate a vision of medical epistemology that requires significant empirical support that is unlikely to be forthcoming any time soon.

References

American Psychiatric Association (APA). 2013. *Diagnostic and Statistical Manual of Mental Disorders: DSM-5*. Washington, DC: American Psychiatric Association.

Aulisio, Mark P. 2003. Meeting the Need: Ethics Consultation in Health Care Today. In *Ethics Consultation: From Theory to Practice*, ed. M. Aulisio, R. Arnold, and S. Youngner, 3–22. Baltimore: Johns Hopkins University Press.

Bechtel, William, and Bob Richardson. 1993. *Discovering Complexity: Decomposition and Localization as Strategies in Scientific Research.* Princeton: Princeton University Press.

Berg, Marc. 1996. Practices of Reading and Writing: The Constitutive Role of the Patient Record in Medical Work. *Sociology of Health & Illness* 18: 499–524.

Boorse, Christopher. 1975. On the Distinction Between Health and Illness. *Philosophy & Public Affairs* 5: 49–68.

———. 1977. Health as a Theoretical Concept. *Philosophy of Science* 44: 542–573.

———. 1997. A Rebuttal on Health. In *What is Disease?* ed. J. Humber and R. Almeder, 3–134. Totowa: Humana Press.

———. 2013. Concepts of Health and Disease. In *Philosophy of Medicine: Handbook of the Philosophy of Science,* ed. Fred Gifford, *vol. 16*, 13–64.

Buchanan, Allen E., and Dan W. Brock. 1989. *Deciding for Others: The Ethics of Surrogate Decision Making*. Cambridge: Cambridge University Press.

Cooper, Rachel. 2002. Disease. *Studies in History and Philosophy of Biology and Biomedical Sciences* 33: 263–282.

Cunningham, Thomas V. (2014). Socializing Medical Practice: A Normative Model of Medical Decision-Making. Doctoral Dissertation, University of Pittsburgh

———. 2015. Objectivity, Scientificity, and the Dualist Epistemology of Medicine. In *Classification, Disease, and Evidence*, ed. P. Huneman, 01–17. Springer Science + Business.

Douglas, Heather E. 2009. *Science, Policy, and the Value-Free Ideal.* Pittsburgh: University of Pittsburgh Press.

Engel, George L. 1977. The Need for a New Medical Model: A Challenge for Biomedicine. *Science* 196:129–136.

Ereshefsky, Mark. 2009. Defining 'Health' and 'Disease'. *Studies in History and Philosophy of Biology and Biomedical Sciences* 40: 221–227.

Frank, Arthur W. 1995. *The Wounded Storyteller: Body, Illness, And Ethics*. Chicago: University of Chicago Press.

Hausman, Daniel M. 2012. Health, Naturalism, and Functional Efficiency. *Philosophy of Science* 79(4): 519–541.

King, Lester. 1954. What is Disease? *Philosophy of Science* 21: 193–203.

Kleinman, Arthur. 1988. *The Illness Narratives: Suffering, Healing, and the Human Condition.* New York: Basic Books.

Lennox, James G. 1995. Health as an Objective Value. *The Journal of Medicine and Philosophy* 20: 499–511.

Machamer, Peter, and Heather Douglas. 1998. How Values Are in Science. *Crit Q* 40: 29–43.

Machamer, Peter, and Gereon Wolters. 2004. Introduction: Science, Values, and Objectivity. In *Science, Values, and Objectivity*, ed. P. Machamer and G. Wolters, 1–13. Pittsburgh: University of Pittsburgh Press.

Machamer, Peter, Lindley Darden, and Carl F. Craver. 2000. Thinking About Mechanisms. *Philosophy of Science*: 1–25.

Margolis, Joseph. 1976. The Concept of Disease. *The Journal of Medicine and Philosophy* 1: 238–255.

Mishler, Elliot G. 1984. *The Discourse of Medicine: Dialectics of Medical Interviews*. Norwood: Alex Publishing Corporation.

Montgomery, K.A. 2008. Sexual Desire Disorders. *Psychiatry (Edgmont)* 5(6): 50–55.

Munson, Ronald. 1981. Why Medicine Cannot Be a Science. *The Journal of Medicine and Philosophy* 6: 183–208.

Quine, W.V.O. 1993. In Praise of Observation Sentences. *The Journal of Philosophy* 90: 107–116.

Schaffner, Kenneth F. 1999. Coming Home to Hume: A Sociobiological Foundation for a Concept of 'Health' and Morality. *The Journal of Medicine and Philosophy* 24: 365–375.

Scheunemann, Leslie P., Robert M. Arnold, and Douglas B. White. 2012. The Facilitated Values History: Helping Surrogates Make Authentic Decisions for Incapacitated Patients with Advanced Illness. *American Journal of Respiratory and Critical Care Medicine* 186: 480–486.

Scheunemann, Leslie P., Thomas V. Cunningham, Robert M. Arnold, Praewpannarai Buddadhumaruk, and Douglas B. White. 2015. How Clinicians Discuss Critically Ill Patients' Preferences and Values With Surrogates: An Empirical Analysis*. *Critical Care Medicine* 43: 757–764.

Simons, M.P., T. Aufenacker, M. Bay-Nielsen, et al. 2009. European Hernia Society Guidelines on the Treatment of Inguinal Hernia in Adult Patients. *Hernia* 13: 343–403.

Sterelny, Kim. 2016. Cooperation, Culture, and Conflict. *British Journal for the Philosophy of Science* 67: 31–58.

Suppes, Patrick. 1977. The Structure of Theories and the Analysis of Data. In *The Structure of Scientific Theories*, ed. F. Suppe, 266–283. Urbana: University of Illinois Press.

Woodward, James. 2003. *Making Things Happen: A Theory of Causal Explanation*. Oxford: Oxford University Press.

Chapter 11
Stochastic Supervenience

Carl F. Craver

Abstract The thesis of physical supervenience (PS) is widely understood and endorsed as the weakest assertion that all facts are tethered to the physical facts. As an exercise in exploring the constitutive commitments of an ontic view of mechanistic explanation, I entertain a weaker tethering relation, stochastic physical supervenience (SPS), the possibility of which is suggested by analogy with the apparent failure of causal determinism (CD) in certain areas of physical science. Considering this possibility helps to clarify the constitutive commitments of mechanistic explanation and to explore the motivations for accepting the PS thesis.

Among the many things I learned from Peter, perhaps chief among them is the power of blurting. This essay is an extended blurt. That said, it's not at all in Peter's style (and barely in my own) but I do hope it does something to honor his intellectual contribution to my and, indeed, our collective understanding of mechanisms. Perhaps if I blurt, someone will feel compelled to respond, and we can pick up on a discussion that Peter started with me long ago about how to understand this puzzling feature of our world.

The norms of mechanistic explanation rest on, and are partly constituted by, a set of commitments about the ontic structure of the world (Haugeland 1998; Kuhn 1962, 106). Most obviously, one seeking a mechanistic explanation is committed to the existence of a causal structure into which explanandum phenomena can be situated and to some understanding of what it means for one thing to "cause" another. In this paper, however, I explore the constitutive commitments of mechanistic explanation with regard to *interlevel* relations: to the relationship between a phenomenon and its underlying mechanism (a constitutive mechanism in the sense of Craver 2007). As a tool for getting at those commitments, I consider the hypothetical possibility that the phenomenon and its underlying mechanism might stand in a stochastic, rather than deterministic, relationship to one another. I show through

C.F. Craver (✉)
Department of Philosophy and PNP Program, Washington University in St. Louis, St. Louis, MO 63130, USA
e-mail: ccraver@wustl.edu

M.P. Adams et al. (eds.), *Eppur si muove: Doing History and Philosophy of Science with Peter Machamer*, The Western Ontario Series in Philosophy of Science 81, DOI 10.1007/978-3-319-52768-0_11

careful consideration of a few simple examples that the possibility of stochastic supervenience relations between a phenomenon and its mechanism (in context) violates the constitutive commitments of mechanistic explanation. This shows that in the constitutive standards of mechanistic explanation, the relationship between a phenomenon and its mechanism is metaphysically stronger than standard supervenience theses express and, in fact, than any nomological (or causal, or merely physically necessary) conception of the relation can supply.

The thesis of physical supervenience (PS) is widely understood and endorsed as the weakest assertion that all facts are tethered to the physical facts. As an exercise in exploring the constitutive commitments of an ontic view of mechanistic explanation, I entertain a weaker tethering relation, stochastic physical supervenience (SPS), the possibility of which is suggested by analogy with the apparent failure of causal determinism (CD) in certain areas of physical science. Considering this possibility helps to clarify the constitutive commitments of mechanistic explanation and to explore the motivations for accepting the PS thesis.

The CD thesis asserts that given a complete specification of the state of the world at time t_0, there is one and only one physically possible state of the world for all $t_n > t_0$. More colloquially, there can be no difference in the state of the world at t_n without a difference in an earlier state of the world. The past facts fix (or determine) all future facts. The CD thesis was once endorsed throughout science and philosophy.

The PS thesis asserts that given a complete specification of the physical state of the world at time t_0, there is one and only one possible total state of the world at time t_0. More colloquially, there can be no difference in any state of the world without there being a difference in the physical state of the world. That is, the physical facts fix (or determine) all the facts. This thesis is now widely accepted by many scientists and philosophers.

The CD thesis, as a general thesis about the causal structure of the world, is now widely believed to be false. To take the most familiar example: there are laws in quantum mechanics that predict, given a complete specification of the antecedent conditions, only probability distributions over possible outcomes. The posit that hidden variables turn these probabilities into certainties entails Bell's inequalities, which are known to be false on the basis of robust experiments (Bell 1964). But one need not appeal to quantum mechanics. CD arguably fails for Newtonian mechanics as well (Alpers et al. 2000). Newton's laws allow objects decelerating from infinite speeds after t_0 to interfere unpredictably with the course of events. And the same laws are consistent with stationary balls sliding spontaneously and unpredictably off the top of a dome in any direction without any motive force (Norton 2003). Regardless of whether one accepts these arguments, it is clear that many scientists and philosophers now question CD, a thesis once held to be beyond empirical reproach.

Might the PS thesis someday suffer a similar fate? Might science discover that the same physical state of the world is consistent with a range of possible total states of the world? I am less interested in answering this question than in exploring what different answers reveal about our metaphysical commitments.

Consider in more detail what SPS would involve. Simplifying matters, imagine a world W containing just three physical magnitudes, A, B, and C, arranged in a causal sequence, $A \rightarrow B \rightarrow C$. Let ϕ be a variable taking different values depending on (i) the values of A, B, and C and (ii) on the nature of the causal relations between A and B and between B and C. For simplicity, assume that A, B, and C are dichotomous such that A can have the values {a, ~a}, B can have the values {b, ~b}, and C can have the values {c, ~c}. And suppose for now that the arrows in the causal sequence stand for necessary and sufficient conditions such that p(b|a) = 1, p(b| ~ a) = 0, p(c|b) = 1, and p(c| ~ b) = 0. Let ψ be a dichotomous variable (with values {ψ_1, ψ_2}) that supervenes on ϕ. If ψ supervenes on ϕ, then there can be no difference in the value of ψ without a difference in the value of ϕ. SPS occurs if a single value for ϕ is consistent with multiple values for ψ. For example, let ϕ_1 = (a→b→c). One simple variety of SPS would allow that in that condition, $p(\psi_1|\phi_1) = 2/3$ and $p(\psi_2|\phi_1) = 1/3$. In more complex cases of SPS, ϕ_1 could be consistent with a probability distribution over arbitrarily many ψ-values {ψ_1, ψ_2, …, ψ_k}. For now, I confine my attention to the simpler, dichotomous case.

Cases of SPS should be contrasted with two other superficially similar kinds of case, each consistent with PS. The first case involves introducing indeterminacy into condition (i). Suppose that A is indeterminate such that p(a) = 2/3 and p(~a) = 1/3. And suppose as above that the arrows in the causal sequence are interpreted as necessary and sufficient conditions. In that case, the probability of ψ_1 given all one knows about W, is 2/3. However, the probability attached to ψ_1 arises from the indeterminate value of A, a component in the supervenience base, and not from any indeterminacy in the supervenience relation itself. That is, the probability of ψ_1 reflects the facts that ϕ has two possible values (ϕ_1, as specified above, and ϕ_2, representing the sequence ~a→ ~ b→ ~ c), that $p(\psi_1|\phi_1) = 1$ and $p(\psi_1|\phi_2) = 0$, and that $p(\phi_1) = 2/3$. In the world we are imagining, once the ϕ-facts are fixed, so are the ψ-facts, even if some of the ϕ-facts are difficult or impossible to fix in advance, as it were.

The second case involves introducing indeterminism into (ii). Suppose that A, B and C take determinate values but that the causal relationship between A and B is stochastic. For example, suppose p(b|a) = 2/3 and p(~b|a) = 1/3, and suppose that the causal relation between B and C is necessary and sufficient as above. In this kind of case, as in the first, the ψ properties would inherit the stochastic features of the supervenience base. But the probability distribution across possible values of ψ does not result from the stochastic nature of the supervenience relation itself. Rather, the probability distribution results from the stochastic nature of the causal relations in the supervenience base. When ϕ takes on the value ϕ_1, as defined above, then $p(\psi_1|\phi_1) = 1$. And when ϕ takes on the value ϕ_3 (where ϕ_3 is the sequence a→ ~ b→~ c), then $p(\psi_1|\phi_3) = 0$. Again, fixing the ϕ-facts fixes the ψ-facts, even if the ϕ-facts do not fix one another. In true cases of SPS as I am constructing it, the stochasticity lies in the supervenience relation itself, not in the supervenience base. (It is surely useful and important to consider how PS might be framed to deal with supervenience relations in causal systems involving indeterminate properties or nondeterministic causal relations; that is not my concern here).

The contrast between SPS and these two contrast cases helps to emphasize that the possibility of SPS in this sense does not arise from imagining how, for example, quantum effects scale up to other levels. Quantum indeterminacy provides merely an analogy and an (at least) apparent air of plausibility for us here. True SPS involves indeterminacy in the supervenience relation itself. To repeat, the thesis of stochastic supervenience would allow that even if the values of A, B, and C are determinate, and even if all of the causal relationships are necessary and sufficient as specified above, it might nonetheless be the case that $p(\psi_1|\phi_1) = 2/3$ and $p(\psi_2|\phi_1) = 1/3$.

The SPS thesis asserts a weaker tethering relationship between a phenomenon and its mechanism than does the PS thesis. Specifically, it allows that some phenomenal facts only stochastically supervene on mechanistic facts. In the context of a broad and weak form of physicalism, SPS might be expressed: All facts at least stochastically supervene on the physical facts. In the context of a given phenomenon and its mechanism, it might be put: All facts about the phenomenon supervene at least stochastically on the facts about its mechanism and the relevant features of its context of operation. Is SPS a genuine metaphysical option?

Some will object to the SPS thesis on the grounds that acknowledging the possibility of SPS amounts to nothing more than the rejection of PS. Yet not all rejections of PS are equal. It is undeniable that if SPS occurs, then the PS thesis is false. However, the PS thesis has always been motivated by the need to express the commitment that non-physical facts are tethered to the physical facts. The SPS thesis can satisfy this motivation as well. Consider again the analogy with apparent failures of causal determinism: Although a single value for a variable representing the antecedent state of the world at t_0 might (if critics of CD are correct) be consistent with more than one effect, the antecedent value might nonetheless be consistent with a finite number of future outcomes or an otherwise manageable range of future outcomes. Indeterminism in causation does not entail that just anything can follow any antecedent condition. At least in many cases considered in the literature, the antecedent value entails a probability distribution over a (perhaps finite) set of future outcomes. By analogy, perhaps the same supervenience base could be consistent with only a finite number of supervenient properties (or a distribution of values of variables describing such properties), and the laws of nature might assign a determinate probability to each. In that case, accepting the possibility of SPS need not lead one to the conclusion that anything goes. Supervenient facts would still be tethered by physical facts; they would simply be more loosely tethered to the physical facts than PS allows. The possibility of SPS need be no more threatening to the search for tethering supervenience relations than the specter of causal indeterminism has been to the search for etiological, causal relations. In fact, as described above, the thesis of global causal determinism just is the assertion that future facts supervene on past facts. If CD is a species of supervenience thesis, then the (at least) apparent failure of CD in some areas of science provides an existence proof for the (at least) apparent possibility of stochastic supervenience relations.

Some (like me) will find the idea of SPS logically incoherent on the grounds that they can find no slack for a supervenient property to differ once its base has been fixed. For those so disposed, SPS can serve as a reminder of how weak the PS thesis really is. As Kim (1998) notes, the PS thesis about the mental is consistent with several varieties of dualism, including psychophysical parallelism and property dualism. If such theses were true, it would be a matter of coincidence, divine providence, or natural law that there is no difference in the mental without there being a difference in the physical. But (on analogy with the failure of CD) there is no reason that these sources of perfect correlation between mind and body could not give rise to a stochastic relation instead. Coincidence places no limits; divine powers are limitless; and (if the above-cited arguments are correct) we already have some reason to believe that some natural laws are stochastic. If one can entertain the possibility of psychophysical parallelism or property dualism, one should have no additional difficulty imagining aspects of the mind and body as standing in stochastic relations to one another.

The SPS thesis is no doubt counterintuitive as applied to standard examples of supervenience in aesthetics and ethics. It seems implausible, for example, that physically identical paintings in physically identical contexts viewed by physically identical observers could be sublime two thirds of the time and banal the rest. And it seems implausible that physically identical agents engaged in physically identical behaviors in physically identical social and political contexts should be praiseworthy two thirds of the time and evil the rest. Examples such as these seem to have none of the nomological or conceptual slack required for SPS.

That said, even if SPS is truly incoherent in aesthetic and ethical contexts, this should not lead one to conclude that SPS is incoherent in all cases. Rather it should lead one to the conclusion that when we imagine such apparently incoherent cases, we are imagining a relation much stronger than mere supervenience (e.g., exhaustive constitution, identity, or conceptual necessity) tethering the supervenient property to its base. The apparent incoherence seems to derive from an underlying presumption that supervenience is underwritten in such cases by a stronger, conceptual or metaphysical relation that precludes the slack required for stochastic supervenience.

The SPS thesis is also counter-intuitive in standard and straightforward cases of mechanistic explanation. Suppose we specify in all relevant detail the entities, activities and organizational features of a standard-model Victor Snaptrap for mice. When it fires (ψ_f), the trigger is depressed, the catch slides, the spring tension releases, and the impact bar arcs toward its target (ϕ_a). Suppose the sequence in ϕ_a is deterministic, that is, p(catch slides|trigger depressed) = 1, p(catch slides|trigger not depressed) = 0, and so on. Here, we find none of the slack required for SPS; if all of the steps in ϕ_a occur, the trap necessarily fires. The trigger was depressed, the catch slid, and the impact bar completed its arc. These steps fully constitute the trap's firing. There's nothing more to be done— no slack for chance. I suppose Chalmers (1996) has something like this in mind when he describes some problems in cognitive science as "easy problems" (relative to the "hard problem" of, e.g., understanding consciousness). In easy problems, we cannot clearly and distinctly conceive of the "higher level" property differing once all of the lower-level steps in the supervenience base have been fixed. One is tempted to say that the higher-level capacity *just is*, or perhaps more perspicuously, *is exhaustively constituted by* the completion of each stage in the operation of the mechanism in its relevant context. Regardless of precisely how one choses to express this relationship in the language of contemporary analytic metaphysics (token identity, exhaustive constitution, realization, or what have you), there is clearly a kind of intimacy our minds cannot render asunder. We cannot clearly and distinctly conceive of ϕ_a being instantiated in its relevant context without firing coming along for free.

Indeed, something like this thought operates as a constitutive ideal in the search for mechanistic explanations (Haugeland 1998). If one knows all of the relevant entities, activities, and organizational features, and if one knows all the relevant features of the mechanism's context of operation, then one knows how the mechanism will behave. If the facts about the parts, activities, organization, and context are determinate and deterministically related, then fixing the ϕ facts fixes the ψ-facts. If, as in our contrast cases, facts about the parts are indeterminate, or if their interactions are indeterministic, then we expect this indeterminacy to be inherited by the probability distributions with which we express the ψ facts. So in attempting to discover how a mechanism works, we take it as an epistemic warning sign when there are significant and salient features of the behavior of the mechanism as a whole (or the probability distributions of its behaviors) that cannot be accounted for (or are incorrectly accounted for) in terms of our understanding of its parts, activities, organization and context. This is because we seem to operate with a background assumption that the phenomenon is exhaustively explained (in an *ontic* rather than epistemic or representational sense; see Craver 2014) by the organized activities of its parts in context.

(An aside: For those who falsely presume that mechanists embrace a "more details better" thesis to the effect that models are always improved by adding details, nothing of the sort follows from what has just been said. One cannot even say what idealization and abstraction *are* without the idea that there is a world, as it is, replete with detail, standing over against our models. If abstraction is dropping detail, there must be detail to drop; if idealizations are deliberate falsities, there must be a truth to willfully abandon. The constitutive ideals of completeness and correctness that underlie and constrain the search for mechanisms does not show up explicitly in our models but rather operates in the metaphysical background, imbuing some models with explanatory significance and giving epistemic force to the distinctive kinds of evidence by which models of mechanisms are evaluated. At any rate, we are talking here about the way things are; not about how we describe them.)

To find the requisite slack for SPS (if we want it), we would need to look beyond standard "easy" cases of successful mechanistic explanation to cases where we can (at least given our current knowledge state) clearly and distinctly imagine the instantiation of the supervenience base without the supervenient property coming along for free.[1] One such case is what Chalmers (1996) calls the "hard problem" of consciousness; perhaps intentionality is another (see Searle 1980). Either due to our ignorance or to some deep fact about the world, we *seem* to be able, clearly and distinctly, to imagine brain mechanisms churning away in their fully specified contexts without any conscious experience or any understanding of what is happening. The real or illusory sense that we can in thought render these phenomena asunder from their mechanisms might open just the kind of slack required for us to imagine SPS.

If one thinks, for example, of "spooky emergence" as a live possibility (i.e., one believes that some properties require but are not exhaustively explained, in the ontic sense, by the causal structure of the physical word) then there is no compelling reason

[1] History is testimony to the fact that many of today's easy problems were yesterday's hard problems: elasticity, nutrition, conception, and heredity. Before these problems were transformed to easy problems one could surely imagine the mechanisms being fixed and the phenomenon not coming along for free.

to favor PS over SPS. Indeed, such a person might find the idea of stochastic supervenience a helpful tool in explicating the idea of spooky emergence. If one thinks of PS in a given domain as holding in virtue of a more metaphysically contingent relation (e.g., as made true by relations of natural law, causation, or divine will), then one should have no difficulty imagining the slack required for SPS. In aesthetics, ethics, and standard mechanistic explanations throughout the sciences, our clear and distinct impressions signal that PS is made true by a stronger kind of metaphysical intimacy than natural law, causation, and divine will can supply. Not even an all-powerful god can make a world where ϕ_a happens and the standard Victor snaptrap doesn't fire.

If one denies the logical coherence of SPS generally, and so fails to find the requisite slack, one is likely embracing a stronger, conceptually or metaphysically necessary view of the relationship between the properties in question than PS describes. On the other hand, if one has no such strong metaphysical commitment, then SPS should be easier to accept than PS. SPS is, after all, a weaker thesis than PS, containing PS as a special case. Why, after all, should those who (unlike this author) embrace the possibility of spooky emergence understand that relation on the model of determinism? If one is willing to untether certain specific properties (e.g., conscious experience) from their mechanistic bases, abandoning a central constitutive commitment of mechanistic science, why then would one tie one's hands by committing to an exceptionally strong metaphysical principle (CD) that even many hard-nosed scientists believe is no longer tenable in the physical sciences?

Even if one grants that the SPS thesis is a live possibility, and even if one acknowledges that it is weaker than the PS thesis and so more worthy of our embrace given the evidence, one might reasonably doubt that SPS ever occurs. I have no evidence that it does. And, as implied above, I find even deterministic forms of e.g., emergence and parallelism unattractive. However, it must be admitted that no empirical evidence, and no compelling metaphysical argument so far given, favors PS over SPS. The fact that several "easy" problems fit the PS thesis has no bearing on how that thesis fairs when we turn our attention to deeper mysteries (concerning, e.g., consciousness and intentionality). PS and SPS should be *at least* equally live options for those inclined to this sort of revisionary metaphysical picture in the first place.

Whether there is any additional reason to embrace SPS remains to be seen. I suspect any such argument will involve inference to the best explanation, i.e., an argument that SPS helps to explain something that would be puzzling if PS were true. After all, that's a standard way of understanding what happened in the above story about Bell's inequalities: the deterministic theory, it was alleged, could not explain the observed probability distributions (as well as an indeterminstic one). Perhaps by introducing slack into the supervenience relationship, one opens new possibilities for seeing some supervenient properties as independent of their supervenience bases and so loosens one central premise in Kim's causal over-determination argument (Kim 1998). If SPS is true in a given case, then the supervenient property can vary independently of the supervenience base, and interventionist theories of causation (for example) might then be used to argue for the causal relevance of higher-level properties. Or another possibility: If (unlike this author) one holds that indeterminism provides "elbow room" for free will, then the concept of stochastic supervenience might help to make space for human freedom in a world of physical causation. Finally, (and to return to the main reason for making these ruminations on stochastic supervenience public) perhaps *the idea* of stochastic supervenience can earn its keep by offering diagnostic leverage in sorting relations that we think hold in virtue of some strong metaphysical necessity (such as identity, exhaustive constitution, or realization) from those that hold in virtue of more contingent relations, grounded in laws, causes, or divine dictates. If one could establish a "no hidden variables" proof for a putative case of stochastic supervenience, then one could conclude that one is dealing with a more contingent kind of relation than identity, etc.

Whether any of these explanatory and diagnostic consequences of considering SPS bear fruit requires more space and time than I can give it here. For now, I merely hope to show that by considering the possibility of SPS we can get a clearer sense of the PS thesis, its metaphysical grounding, and its role as a constitutive ideal in mechanistic sciences. Along the way, we have also found some new reasons to avoid assimilating interlevel relations in a hierarchy of mechanisms to causal relations and, accordingly describing a constitutive mechanism as producing (rather than exhaustively constituting in context) the phenomenon it underlies. I hope that by blurting a few commitments and following them out, this essay offers a novel anchor point for (or entry to the discussion of) the ontological relations between the multiple causal levels of our complex world.

References

Alpers, J.S., M. Bridger, J. Earman, and J. Norton. 2000. What Is a Newtonian System? The Failure of Energy Conservation and Determinism in Supertasks. *Synthese* 124: 281–293.

Bell, J.S. 1964. On the Einstein-Podalsky-Rosen Paradox. *Physics* 1: 195–200.

Chalmers, D. 1996. *The Conscious Mind*. Oxford: Oxford University Press.

Craver, C.F. 2007. *Explaining the Brain: Mechanisms and the Mosaic Unity of Neuroscience*. Oxford: Clarendon Press.

———. 2014. The Ontic Conception of Scientific Explanation. In *Explanation in the Biological and Historical* Sciences, eds Andreas Hütteman and Marie Kaiser. Dordrecht: Springer.

Haugeland, J. (1998). *Having thought*. Harvard University Press.

Kim, J. 1998. *Mind in a Physical World*. Cambridge, MA: MIT Press.

Kuhn, T. 1962. *The Structure of Scientific Revolutions*. Chicago: University of Chicago Press.

Norton, J.. 2003. Causation as Folk Science. *Philosophers Imprint*. www. Philosphersimprint. org/003004

Searle, J. 1980. Minds, Brains, and Programs. *The Behavioral and Brain Sciences* 3: 417–457.

Chapter 12
Activities Are Manifestations of Causal Powers

Gualtiero Piccinini

Abstract I reduce activities to properties, where properties include causal powers. Activities are manifestations of causal powers. Activities occur when an entity's causal powers encounter partners for their manifestation. Given this reduction of activities to (manifestations of) properties, entities and properties (and their organization) are all we need for an ontology of mechanisms.

What are mechanisms? Machamer, Darden, and Craver (MDC) say this:

> Mechanisms are composed of both *entities* (with their properties) and *activities*. Activities are the producers of change. Entities are the things that engage in activities. Activities usually require that entities have certain types of properties" (MDC 2000, 3; cf. Machamer 2004).

By listing activities separately from entities and their properties, MDC imply that activities are not reducible to properties. Later in the MDC paper (Sect. 12.3) as well as in a separate paper (Machamer 2004), Peter Machamer explicitly rejects attempts to reduce activities to entities and their properties. According to Machamer, our ontology must include activities *in addition* to entities and their properties.

But in that same paragraph above, MDC also state that activities "usually require" that entities have certain properties. This suggests a tight connection between activities and properties, which MDC do not explicate further—except by providing a few examples of how activities seem to depend on properties. This paper investigates the connection between activities and properties.

MDC have several reasons for positing activities as a separate category. One is that activities account for change in a causal way that fits the scientific practices of neurobiology and molecular biology—as opposed to laws of nature, which are supposed to account for change but play almost no role in those sciences (MDC, 7–8). Another reason is that, Machamer (2004) argues, activities are ontologically prior to causal powers. Finally, sometimes MDC appear to think of properties along nominalistic lines, and they seem to assume that nominalistically understood properties

G. Piccinini (✉)
Philosophy Department, University of Missouri-St. Louis, St. Louis, MO 63121, USA
e-mail: piccininig@umsl.edu

M.P. Adams et al. (eds.), *Eppur si muove: Doing History and Philosophy of Science with Peter Machamer*, The Western Ontario Series in Philosophy of Science 81, DOI 10.1007/978-3-319-52768-0_12

are insufficient to account for activities. But they don't discuss the different options in the metaphysics of properties explicitly. (I fill this lacuna below.)

MDC's dualism of entities and activities has met mixed reactions. Some see it as an important innovation (e.g., Illari and Williamson 2013; Glennan forthcoming, Chap. 2). Others are less sanguine and prefer to avoid committing themselves to activities as an irreducible category; perhaps because of this, when formulating their account of mechanisms, they avoid mentioning activities and talk of functions or capacities instead (e.g., Psillos 2004; Bechtel and Abrahamsen 2005). But functions or capacities raise the same issue—are they an irreducible category or are they reducible to entities and their properties?

The ontology of science lists the kinds of thing we invoke in our scientific practices—our basic categories—and determine whether some are reducible to others. For example, Hume argued that all there is to causation is constant conjunction between events. I disagree with Hume's putative reduction of causation. But I endorse the project of finding a minimal list of categories, to which everything else reduces. This is the sense in which MDC maintain that activities are irreducible to entities and their properties. From now on, when I write 'reduction' I mean reduction in this metaphysical sense, not to be confused with reduction between levels of reality or between scientific theories.

Let's agree with Machamer that we need to posit entities and their properties. Do we need to add activities (or functions, or capacities) as an additional category, as Machamer argues, or can we account for activities in terms of entities and their properties?

I will argue that there are two ways to reduce activities to properties, which depend on whether properties include causal powers. On any account that keeps causal powers out of properties—whether it be a nominalist or realist account—activities may be seen as manifestations of properties plus laws of nature. This account does not sit well with mechanists like Machamer, so mechanists may prefer to posit activities as a separate category (instead of appealing to laws).

By contrast, on any account that sees causal powers as inherent in properties, activities are manifestations of properties alone—without help from laws of nature. Activities are manifestations of causal powers. Activities occur when an entity's causal powers encounter partners appropriate for their manifestation. This reduces activities to (manifestations of) properties.

I will argue that activities thus understood can do all the jobs that mechanists need them to do and that the underlying ontology of properties as including causal powers is better suited to the mechanistic framework than an ontology of powerless properties plus laws of nature. Therefore, there is no reason to posit activities *in addition* to properties. By being reduced to (manifestations of) properties, activities find a natural and satisfying place in our ontology without inflating it.

In the next section, I will look at accounts of mechanisms that appeal to functions or capacities (rather than activities) and how they relate to accounts of mechanisms that appeal to activities. In the following section, I will look at the relationship between properties and causal powers. This paves the way for arguing, in the last section, that activities are manifestations of causal powers.

12.1 Activities Versus Functions

When I define mechanisms, I usually do so in terms of components performing functions rather than activities (e.g., Piccinini 2015). I'm not alone. For example, Bechtel and Abrahamsen define a mechanism as "a structure performing a function in virtue of its component parts, component operations, and their organization" (2005, 423). Since components of mechanisms are often mechanisms, it follows from Bechtel and Abrahamsen's characterization that components of mechanisms perform functions. Are accounts of mechanisms in terms of functions different from accounts in terms of activities? How are functions and activities related?

Functions can be understood in two main ways. One is teleological, the other is not. In the nonteleological sense, 'performing a function' means something very close to 'performing an activity.' Therefore, defining mechanisms in terms of performing nonteleological functions is very close to defining mechanisms in terms of performing activities.

The one possible difference is that a function in the nonteleological sense is sometimes defined as an activity that is relevant from a perspective that we take towards what a system does (Wimsatt 1972). According to this view of functions, functions are perspectival in a way that activities per se may not be. An especially important perspective is the perspective from which we look at the contribution of a function to the capacity of a containing system. This perspective is especially important because it allows us to explain the functions or capacities of a system in terms of its subfunctions or subcapacities. This explanatory strategy is the basis of functional analysis as well as mechanistic explanation. From this perspective, functions may be seen simply as contributions to the capacities of a containing system (Cummins 1975; Craver 2012).

Capacities and subcapacities *just are* causal powers—or so I assume. In subsequent sections, I will argue that activities are manifestations of causal powers. If so, then activities are manifestations of capacities as well. For now, let's grant that the manifestations of capacities are activities. If a (nonteleological) function is a capacity and the manifestations of capacities are activities, then the performance of a (nonteleological) function is an activity.

Presumably, for any activity there are perspectives from which that activity is relevant to what a system does. This is especially salient if the perspective in question is mechanistic. Mechanistic explanations that appeal to activities explain the activities of whole mechanisms in terms of subactivities. This works in exactly the same way as mechanistic explanations that appeal to nonteleological functions or capacities, which explain the functions or capacities of whole mechanisms in terms of subfunctions or subcapacities. Therefore, any activity is the performance of a (nonteleological) function from some mechanistic perspective. Thus, there is no relevant difference between accounts of mechanisms that appeal to nonteleological functions and accounts that appeal to activities. Even if functions are perspectival, defining mechanisms in terms of nonteleological functions is roughly equivalent to defining them in terms of activities. Therefore, the discussion of activities in the following sections applies to nonteleological functions too.

By contrast, in the teleological sense of 'function', not all activities count as functions. In the teleological sense, we say that the function of a flashlight is to make light, not to consume batteries, even though flashlights do both. We say that a function of the nose is letting air into the respiratory system, not letting germs in, even though noses do both. In the teleological sense, entities that fail to perform their functions at the appropriate rate are malfunctioning or defective. In other words, performing a teleological function is a special sort of activity—teleological functions have something that activities, in general, may lack.

Corresponding to the distinction between teleological and nonteleological functions, we may distinguish between mechanisms that perform activities (i.e., nonteleological functions) in general, and *functional* mechanisms, which perform at least some teleological functions in addition to other activities (cf. Garson 2013). Functional mechanisms are a species of mechanisms in general, a species which includes typical biological mechanisms, typical artifacts, and their components.

Corey Maley, Justin Garson, and I have argued that teleological functions, in the sense relevant to most special sciences, reduce to complicated sets of causal powers, their manifestations, and relations between them. Briefly, Maley and I (forthcoming) argue that teleological functions are stable causal contributions towards the goals of organisms, by either a trait or an artifact of the organisms. Goals, in turn, divide into two classes. Objective goals are outcomes that organisms must pursue on pain of extinction—namely, survival and inclusive fitness. Subjective goals are outcomes that organisms can pursue in virtue of being sapient and sentient, and they may or may not be in conflict with objective goals. Garson and Piccinini (2014) offer a detailed account of the conditions under which biological functions—those that contribute to survival and inclusive fitness—are performed at appropriate rates in appropriate situations. More work needs to be done to fully develop this account. The important point is that the account, if successful, reduces teleological functions to current causal powers.

As I pointed out at the beginning of this section, nonteleological functions just are activities. If we could also show that (i) activities reduce to causal powers and (ii) causal powers are aspects of properties, we would obtain the nice result of reducing all functions—teleological and nonteleological—to aspects of properties. This means that we could do all the work needed for an ontology of mechanisms, whether or not they perform teleological functions, in terms of entities and properties (and their organizational relations).

12.2 Properties and Causal Powers

To set the stage for whether activities reduce to properties, I will now look at different accounts of properties and what they make of the relation between properties and causal powers. I assume there are entities—what metaphysicians typically call particulars or substances. I also assume there are properties. But there are many accounts of properties and they make a difference to whether and how activities reduce to properties.

According to realism (as I use the term), properties are real in their own right. According to nominalism (as I use the term), properties are reducible to concrete particulars. For example, Class Nominalism is the view that properties are classes of particulars (e.g., Lewis 1983); Resemblance Nominalism is the view that for a particular to have a property is to resemble other particulars (Rodriguez-Pereyra 2002).

If nominalism about properties is right, properties alone cannot account for activities. For activities produce change (MDC 2000, 3; Machamer 2004, 29). But how can classes of particulars or resemblances between particulars, by themselves, produce any change? Put differently, how can an entity produce change simply in virtue of belonging to a class of entities or resembling some other entity?

Consider the mass *m* of a planet orbiting a star. According to Class Nominalism, all there is to having mass *m* is belonging to a specific class of entities; according to Resemblance Nominalism, all there is to having mass *m* is to resemble certain other entities (entities of mass *m*). But belonging to one class rather than another, or resembling certain entities rather than others, by itself, produces no change at all, let alone the specific change of causing a planet to orbit its star.

Because of this, nominalistically understood properties need help to produce change. Such help is usually presumed to come from laws: nominalists posit laws in addition to particulars, so that properties plus laws can be said to determine change. Change is the shift of particulars from one class to another, or the shift of which other particulars a particular resembles, and such shifts are brought about by laws. In our example, it is the law of gravitation that, in addition to a planet's property of having mass *m*, ensures that the planet shifts from one position along its orbit to the next.

So a nominalist might propose to account for activities in terms either of classes of particulars or resemblances between particulars—plus laws. An activity is what happens when a particular falls under a law, in virtue of belonging to a certain class or by resembling certain other particulars.

This reduction of activities to particulars plus laws runs into a problem. As Machamer and other mechanists often point out, the special sciences—sciences that dwell in mechanisms—typically have little use for laws. They discover few if any laws and they don't explain phenomena in terms of laws (e.g., Kim 2012). Instead, the special sciences explain phenomena in terms of mechanisms. For a metaphysician to then posit laws that explain the activities posited in mechanistic explanations would be at odds with scientific practices. In fact, MDC present mechanistic explanation as an alternative to the traditional account of explanation in terms of laws of nature: reducing mechanisms to particulars plus laws is a coherent position, but it devolves into the traditional account of change and explanation that mechanists were trying to move away from.

Perhaps we can get around at least the first part of this problem—that special sciences have little use for laws—by reducing all activities to fundamental microphysical particulars plus fundamental laws. For it is relatively uncontroversial that at least fundamental physics discovers fundamental laws and explains phenomena in terms of such fundamental laws. So here's a possible nominalistic account of activities, and of mechanisms more generally. Imagine the fundamental particulars—the fermions and bosons of the standard model, perhaps. Depending on which class they belong in, or which other particulars they resemble, fundamental particulars fall under fundamental laws. Based on which fundamental laws they fall under, they act in certain ways—they engage in specific activities.

Now compose and organize the particles into complex particulars. Such complex particulars will engage in specific activities that emerge from the activities of their constituents and the way their constituents are organized. Now continue composing and organizing complex particulars into more complex particulars, level by level, until you constitute all levels of interest (atoms, molecules, tissues, organs, organisms, etc.). At each level, the organized activities of the components give rise to the activities of the whole. But all activities boil down to complex arrangements of fundamental particles governed by fundamental laws. Thus, mechanisms and their activities reduce to fundamental particulars plus fundamental laws and organizational relations.[1]

Although this nominalistic account is both principled and elegant, it is unattractive from a mechanistic perspective. Mechanists like Machamer insist that mechanisms have a sort of *agency*, and that's why mechanists posit activities. But the nominalistic reduction takes all active oomph away from particulars and gives it to laws. If activities are ultimately reduced to particulars falling under laws, the agency ultimately belongs in the laws, not in the particulars. In this sense, the mechanisms' agency turns out to be illusory. The deep story about mechanisms is that their agency derives from the fundamental laws of nature.[2] In addition, the second part of our original problem remains. According to mechanists, explanation is provided by mechanisms, not by laws.

At this point, one option is to posit activities as their own categories, as Machamer does. Instead of laws, activities are posited in addition to properties to get particulars to act. Depending on which class a particular belongs to, or which other particulars it resembles, it engages in specific activities and not others. Instead of laws producing change, activities do.

Another option is to see if realism about properties gives us a mechanistically more attractive alternative than nominalism. According to realism about properties, properties may be either universals (e.g., Armstrong 1997) or abstract particulars (a.k.a. tropes, or modes; e.g., Heil 2012). I will remain neutral on this point, or at

[1] As usual, this is reduction in the sense that the category of activities reduces to the categories of particulars and laws; it is not a reduction in the more popular sense that higher level properties are identical to lower level ones. I am staying neutral as to whether the reduction of activities to fundamental particulars and laws involves identities between higher levels and lower levels or some other relation such as realization.

[2] Unless laws are just regularities, in which case the deep story is that agency itself is illusory.

least neutral between Aristotelian universals (in re universals) and abstract particulars. (I reject abstract objects, so I reject Platonic universals.)

For present purposes, more important than whether properties are universals or particulars is the relation between properties and causal powers. There are three main views: (i) properties are purely qualitative or categorical (*categoricalism*; e.g., Lewis 1986), (ii) properties are sets of causal powers or dispositions (*dispositionalism*; e.g., Shoemaker 1980; Bird 2007), and (iii) properties are both qualities and powers because qualities and powers are the same thing (*identity theory*; Martin 2008; Heil 2003, 2012).

Options (ii) and (iii) have the virtue of tying powers directly to properties, so that properties just are powers. The first option does not have this virtue. In this respect, pure qualities are like classes of particulars or resemblances between particulars: on their own, they are inert. They need something else in order to produce change—presumably laws. Laws again! Like a nominalist, a categoricalist may suggest helping herself to laws, so that qualities plus laws produce change.

This realist route to reducing activities to properties has similar pros and cons to the nominalistic route we just explored. As before, I set this route aside because it defeats the purpose of shifting from a nomological account of change and explanation to a mechanistic account. As before, one option at this point is to posit activities as their own category, instead of laws. That is Machamer's preferred option. But there is another option: consider the possibility that properties include causal powers.

As we've seen, there are two ways of linking properties and powers. Either properties are pure powers or properties are both categorical and dispositional at the same time because qualities and powers are the same thing. For present purposes, either of these options is acceptable. What matters is that, on either of them, powers stem directly from properties.

Before proceeding, it's important to note that powers manifest themselves under appropriate conditions. Different conditions lead to different manifestations of the same powers. Consider the power of a pencil to leave marks on paper. If a pencil is pressed against paper very lightly (condition C1), it leaves no mark (manifestation M1). If it is pressed lightly (condition C2), it leaves a faint mark (manifestation M2). If it is pressed more heavily (condition C3), it leaves a heavier mark (manifestation M3). And so on.

Another way to put the point is that dispositions (powers) manifest themselves in the presence of appropriate partners and manifest themselves differently in the presence of different partners (cf. Martin 2008). Partners for a power p of a given particular are powers belonging to other particulars that bring about p's manifestations, while p, in turn, brings about their manifestations. Consider our pencil again. If nothing is pressing the pencil against paper (presence of paper, which is one partner, but lack of another partner, namely something holding the pencil and pressing it against the paper), it leaves no mark. If someone presses the pencil against paper too lightly (lack of appropriate partner), it still leaves no mark. If someone presses the pencil against paper more forcefully (presences of appropriate partner), it leaves a mark on the paper that depends on how hard it is pressed. With these preliminaries in place, we can now reduce activities to manifestations of powers.

12.3 Activities Are Manifestations of Powers

I propose that activities *just are* manifestations of causal powers (dispositions), and vice versa. Leaving marks on paper is a manifestation of the pencil's power under appropriate conditions. Equivalently, leaving marks on paper is a manifestation of the pencil's power when the pencil encounters appropriate partners.

For a more mechanistic example, consider an air conditioner's activity of cooling the inside of a house. It is a manifestation of the air conditioner's cooling power. Such cooling power is not always manifested. It is manifested when the air conditioner is in working order, appropriately located relative to a house, plugged in, turned on, air is present in the environment, and electricity is running through the electrical circuit. The air conditioner needs appropriate partners—the boundaries and insulation provided by the house, electricity, air, something turning it on—in order to manifest its cooling power. When it meets appropriate partners, it cools—it manifests its cooling power by actually cooling the house.

A mechanistic explanation explains the activities of the whole in terms of the organized activities of the component parts. In our example, the components of air conditioners have specific powers (compressing gases, blowing gases, etc.) that, when organized in the appropriate way to form the air conditioner and when the conditions are right (i.e., when they meet their appropriate partners, such as electrical energy, refrigerant in the pipes, air in the environment, etc.), manifest themselves by producing a number of activities (compressing refrigerant, blowing air, etc.) that together constitute the activity of cooling the house.

In general, any activity is the manifestation of one or more causal powers; any manifestation of a causal power is an activity. To establish the adequacy of this account, I will go through what Machamer says about the ontology of activities and argue that manifestations of causal powers have all the necessary requisites.

The most important feature of activities is that they produce change (MDC 2000, 3; Machamer 2004, 29). Causal powers are perfectly suited to account for change. Causal powers are powers to cause effects; one type of effect is to produce change. Causal powers (sometimes) bring about change when they are manifested. Another type of effect is to maintain the status quo—many mechanisms preserve the status quo rather than produce change. For example, bridges, buildings, and tables are built to support things; their main activity is to preserve themselves and what they support as they are. This means that activities themselves do not necessarily produce change; sometimes activities preserve the status quo.

According to the present account, then, powers manifest themselves if and only if activities occur, because activities are manifestations of causal powers. Causal powers, in turn, are an aspect of properties. Thus, the manifestations of causal powers (properties) are perfect candidates for activities qua producers of change or preservers of the status quo.

Machamer suggests that activities belong to entities in a different way than the way "structural properties" belong to entities: "activities are how entities express themselves. Activities show how entities are not just static (or passive) and in what ways they act as causal agents" (2004, 29). These features of activities fit the view I advocate like a glove. Activities are how entities express themselves, indeed. Entities express themselves by manifesting their causal powers; activities are just manifestations (i.e., expressions) of their causal powers. As mentioned before, this shows how entities are not just passive but also active. Nothing could be better suited to showing this than the entities' causal powers.

But Machamer goes beyond the above point:

> [I]t is not clear even that all activities are necessarily the activity of some entity, or, less strongly, that one always can or needs to identify an entity to which an activity belongs. It is unclear to me that forces, fields, or energy are entities and not, as it were, "substantial" activities, or that the process or activities of equilibrating or reaching stasis need entities in order to be understood (Machamer 29–30).

Here Machamer is giving examples of activities that may be "substantial" in the sense of existing independently of any entities. The different examples require different treatments. Forces are activities and they belong to entities—they are manifestations of the causal powers of the entities that exert the forces. Energy is not an activity but a causal power—it belongs to the entities that possess the energy and may be transferred from one entity to another. Fields are either causal powers of spacetime itself or entities in their own right. Either way, there is no need to posit "substantial" activities in addition to entities and their causal powers, whose manifestations are the activities. But if there were "substantial" activities, I propose that they would be manifestations of "substantial" causal powers—namely, causal powers that exist independently of any entities. Finally, if some activities—such as equilibrating or reaching stasis—don't need entities in order to be understood, then some manifestations of some causal powers don't need entities in order to be understood. In the end, we can account for Machamer's examples in terms of causal powers and their manifestations.

Machamer also argues that activities can be abstracted away from entities and therefore can be referred to and identified independently of entities (2004, 30). I agree. Since the same is true of causal powers, this is ok. From this, though, Machamer concludes that "activities existing as abstract objects exist independently" (ibid.). This is a non sequitur. From the epistemic point that activities (i.e., manifestations of causal powers) can be considered in abstraction from entities, it doesn't follow that activities are abstract objects that exist independently of entities.

Maintaining that activities exist as abstract objects amounts to platonism with respect to activities. This is one possible view, and it can be combined with a version of the present account that relies on platonism about properties. The result is that abstract activities are (abstract?) manifestations of abstract causal powers. I don't know what the (abstract?) manifestation of abstract causal powers is or how it works. But then again, I don't understand what any abstract entities are or how they work, not to mention how we can know about or refer to them (Benacerraf 1973; Field 1989; Balaguer 1998). Since I don't believe in abstract entities at all, I will set abstract entities aside.

Machamer's strongest objection to reducing activities to (manifestations of) powers is that activities are "ontologically prior" to powers (2004, 30). He gives three reasons. First, our concepts for powers presuppose our concepts for the activities exhibited by manifesting those powers. Second, we can't define powers without referring to the activities exhibited by manifesting those powers. Third, we must experience activities before we can understand powers (ibid.; cf. Illari and Williamson 2013, §3). For example, our concept of fragility presupposes the concept of breaking, we can't define fragility without referring to the breaking of things, and we can't understand what fragility is without having experienced the breaking of something.

It may well be that activities are *epistemically* prior to powers in the sense that our concepts of activities presuppose our concepts of powers, our definitions of powers require reference to activities, and activities must be experienced in order to understand powers. Whether and to what degree this is the case is a very difficult question, which I will not take up here. I will not take it up because what is in contention is whether activities are *ontologically* prior to powers, and that is a separate question. From the epistemic priority of activities in the sense just described it doesn't follow that activities have any ontological priority. There is a compelling reason to conclude the opposite.

Powers are open-ended in a way that outstrips their manifestations, and hence the activities that are their manifestations (cf. Martin 2008). Any child is familiar with this. Children play with toys, random objects, and even their own hands and feet. Both children and adults, including scientists, explore what happens to objects under various conditions and manipulations. At any given time, there are indefinitely many manipulations that have not been performed or conditions that remain unobserved. Yet as soon as a few manipulations are performed, we infer that objects have powers, which will give rise to as yet unobserved activities under further manipulations or conditions. After enough manipulating and observing, we start predicting what will happen under what circumstances. Turn the object one way and it will look one way. Turn it another way and it will look another way. Bang it one way and it will make one noise. Bang it another way and it will make a different noise. And so forth.

After we observe enough activities under enough conditions, we observe that activities are modulated by various conditions and, even better, we observe how activities are modulated by varying the conditions ourselves. Eventually, we become pretty good at predicting how new variations in the conditions will vary activities, and we eventually infer that entities have powers that go beyond any particular activity they manifest. There are indefinitely many ways an object can look or can make noises, most of which will never occur. That is to say, objects have powers whose potentiality goes beyond their actual manifestations. We infer powers from activities precisely because we realize that things *can* do more than what they *actually* do. Similarly, we eventually learn from our own case that we *can* do all sorts of things, beyond what we've already done. What we can do are our powers.

Because powers are open-ended in a way that activities are not, powers cannot be reduced to activities and activities are not ontologically prior to powers. On the contrary, powers give rise to activities by manifesting themselves under relevant circumstances. Therefore, powers are ontologically prior to activities and activities reduce to (manifestations of) powers.

12.4 Conclusion

Machamer is right that activities cannot reduce to properties—under some accounts of properties. Specifically, activities cannot reduce to properties alone under any account that separates properties from causal powers. If properties do not involve causal powers directly (e.g., if there aren't any causal powers), it may still be possible to reduce activities to something else, but it won't be to properties alone. Activities may reduce to properties plus the laws of nature. This is not a mechanistically appealing account of activities, because mechanists argue that explanation is provided by mechanisms not laws.

But under any account of properties that includes causal powers as aspects of properties, we can reduce activities to properties in a way that does justice to the scientific practices that motivate mechanists. This is good because, once we posit entities and their properties (including their organizational relations), we already have everything we need for our account of mechanisms. We don't need to posit activities in addition to properties: activities are manifestations of powers, which occur when the properties of entities encounter appropriate partners.

Acknowledgement Thanks to Carl Craver and John Heil for helpful discussions on this topic, and to Billy Dunaway, Uljana Feest, John Heil, Lauren Olin, and Jacqueline Sullivan for helpful comments on previous drafts.

References

Armstrong, D.M. 1997. *A World of States of Affairs*. Cambridge: Cambridge University Press.

Balaguer, M. 1998. *Platonism and Anti-Platonism in Mathematics*. Oxford: Oxford University Press.

Bechtel, W., and A. Abrahamsen. 2005. Explanation: A Mechanistic Alternative. *Studies in History and Philosophy of the Biological and Biomedical Sciences* 36: 421–441.

Benacerraf, P. 1973. Mathematical Truth. *The Journal of Philosophy* 70: 661–679.

Bird, A. 2007. *Nature's Metaphysics: Laws and Properties*. Oxford: Oxford University Press.

Craver, C.F. 2012. Functions and Mechanisms: A Perspectivalist Account. In *Functions*, ed. P. Huneman. Dordrecht: Springer.

Cummins, R. 1975. Functional Analysis. *The Journal of Philosophy* 72(20): 741–765.

Field, H. 1989. *Realism, Mathematics, and Modality*. New York: Basil Blackwell.
Garson, J. 2013. The Functional Sense of Mechanism. *Philosophy of Science* 80: 317–333.
Garson, J., and G. Piccinini. 2014. Functions Must Be Performed at Appropriate Rates in Appropriate Situations. *British Journal for the Philosophy of Science* 65(1): 1–20.
Garson, J., forthcoming. *The New Mechanical Philosophy*. Oxford: Oxford University Press.
Heil, J. 2003. *From an Ontological Point of View*. Oxford: Clarendon Press.
——. 2012. *The Universe as We Find It*. Oxford: Clarendon Press.
Illari, P., and J. Williamson. 2013. In Defence of Activities. *Journal for General Philosophy of Science* 44(1): 69–83.
Kim, J. 2012. Against Laws in the Special Sciences. *Journal of Philosophical Research* 37: 103–122.
Lewis, D. 1983. New Work for a Theory of Universals. *Australasian Journal of Philosophy* 61: 343–377.
———. 1986. *On the Plurality of Worlds*. Oxford: Basil Blackwell.
Machamer, P. 2004. Activities and Causation: The Metaphysics and Epistemology of Mechanisms. *International Studies in the Philosophy of Science* 187(1): 27–39.
Machamer, P., L. Darden, and C.F. Craver. 2000. Thinking About Mechanisms. *Philosophy of Science* 67(1): 1–25.
Maley, C.J., and G. Piccinini. forthcoming. A Unified Mechanistic Account of Teleological Functions for Psychology and Neuroscience. In *Integrating Mind and Brain Science: Mechanistic Perspectives and Beyond*, ed. David Kaplan. Oxford: Oxford University Press.
Martin, C.B. 2008. *The Mind in Nature*. Oxford: Oxford University Press.
Piccinini, G. 2015. *Physical Computation: A Mechanistic Account*. Oxford: Oxford University Press.
Psillos, S. 2004. A Glimpse of the Secret Connexion: Harmonizing Mechanisms with Counterfactuals. *Perspectives on Science* 12(3): 288–319.
Rodriguez-Pereyra, G. 2002. *Resemblance Nominalism. A Solution to the Problem of Universals*. Oxford: Clarendon Press.
Shoemaker, S. 1980. Causality and Properties. In *Time and Cause: Essays Presented to Richard Taylor*, ed. P. van Inwagen, 109–135. Dordrecht: Reidel.
Wimsatt, W.C. 1972. Teleology and the Logical Structure of Function Statements. *Studies in History and Philosophy of Science* 3: 1–80.

Chapter 13
Back to the Cradle: Mechanism Schemata from Piaget to DNA

Catherine Stinson

Abstract Mechanism schemata are one of the least understood parts of MDC's account of mechanistic explanation. Relatedly, there is a common misconception that there is no place for abstraction in MDC mechanisms. These two problems can be remedied by looking more carefully at what MDC say both in their 2000 paper and elsewhere about schemata and abstraction, and by following up on a comment of Machamer's indicating that Piaget was the inspiration for schemata. Darden's work on mechanism discovery reveals an important role for abstract mechanism schemata, not only as concise representations of known mechanisms, but also as hypotheses for how unknown mechanisms might work. Piaget's schemata likewise both capture the developing child's current state of knowledge, and serve as templates into which new experiences are fit. If we assume that mechanistic explanations should pick out real difference-makers, and that what makes a difference needn't always be fine-grained details, we open up an explanatory role for abstract schemata to play. Schemata represent the coarser-grained features of mechanisms that are often among the most important difference-makers. In many contexts they might need to be supplemented with finer grained details in order to give an adequate explanation, but independent of those details, schemata remain explanatory in virtue of being part of the explanation. Detailed accounts of real difference-makers need not be seen as in competition with abstract models or generalizations. They can (and should) be combined in mechanistic explanations.

13.1 Introduction

While writing my dissertation under the joint supervision of Peter Machamer and Ken Schaffner, there were several occasions when Peter and I discovered that we had diametrically opposed views. Once or twice Ken quietly assured me after the end of the meeting that he agreed with me. But more often I would spend the next

C. Stinson (✉)
Rotman Institute of Philosophy, University of Western Ontario, London, ON, Canada
e-mail: cstinso5@uwo.ca

M.P. Adams et al. (eds.), *Eppur si muove: Doing History and Philosophy of Science with Peter Machamer*, The Western Ontario Series in Philosophy of Science 81, DOI 10.1007/978-3-319-52768-0_13

several months or years trying to figure out how Peter could possibly be right about something as obviously wrong as, for example, top-down causation.

One thing that we largely agreed on was the role of abstraction in mechanisms, although we seemed to disagree with nearly everyone else. We did, however, disagree on how to pluralize "schema," which he told me had also been a bone of contention during the writing of (Machamer et al. 2000) [hereafter MDC] too. Peter was for 'schemata.' Lindley Darden was for 'schemas.' Peter prevailed. I suspect the disagreement over spelling belies a deeper ambiguity, which is manifest in MDC's unclarity about what a schema is.[1]

At this point in the conversation Peter switched from his usual booming voice to the softer one he uses when he's sharing deeply held beliefs, to tell me that what he had in mind for schemata was Piaget. I nodded in silent understanding, although I had no idea what he was talking about. Several years later I've finally gotten around to looking it up. Here I will use Piaget's schemata to help clarify some misunderstandings of mechanism schemata, and to show how abstraction might fit into the MDC view of mechanistic explanation.

I begin by looking at how schemata are described in MDC, how they are understood by subsequent commentators, and how they are described in Darden's work. Next I introduce Piaget's schemata and draw out the characteristics of those that Peter had in mind, in order to elaborate how mechanism schemata are meant to capture abstract difference-makers with broad applicability. I contrast that with the popular reception of the MDC view, where Craver in particular is taken to speak against there being an explanatory role for abstract mechanisms. Craver's focus on detail in causal mechanisms seems to preclude the essential role that schemata play not only in mechanism discovery, but also in explanation. I then describe a way of interpreting MDC that makes room for both detail and abstraction, and that is more consistent with the textual evidence than the popular interpretation.

13.2 Mechanism Schemata

The MDC characterization of mechanism says, "Mechanisms are entities and activities organized such that they are productive of regular changes from start or set-up to finish or termination conditions" (Machamer et al. 2000, 3). Part of the paper that has Peter's stamp on it is the focus on types of mechanisms: "Scientists... are typically interested in types of mechanisms, not all the details needed to describe a specific instance of a mechanism. We introduce the term 'mechanism schema' for an abstract description of a type of mechanism" (Machamer et al. 2000, 15). In this passage, schemata sound something like templates or abstract models, and it is noted that the mechanisms scientists are interested in are typically not highly detailed.

[1] This makes one wonder whether there might be deep messages hidden in some of Peter's other idiosyncratic spelling preferences.

The very next sentence is more often quoted though: "A *mechanism schema* is a truncated abstract description of a mechanism that can be filled with descriptions of known component parts and activities" (Machamer et al. 2000, 15). This is often paired with with the definition of a mechanism sketch: "For epistemic purposes, a *mechanism sketch* may be contrasted with a schema. A sketch is an abstraction for which bottom out entities and activities cannot (yet) be supplied or which contains gaps in its stages" (Machamer et al. 2000, 18). The contrast between mechanism schemata having known components while the entities of a sketch are not yet known suggests that the difference is just a matter of the state of our knowledge about the mechanism. The case study MDC present of discovering the mechanisms of protein synthesis places the schema of the central dogma of molecular biology as a later stage of mechanism discovery after Watson's earlier sketch, which may also lend this impression. Craver later describes schemata in just this way. He says, "Between sketches and complete descriptions lies a continuum of *mechanism schemata* whose working is only partially understood" (Craver 2007, 114). This is one of the only passages in Craver's book where he mentions schemata at all. Many commentators have understandably followed Craver in interpreting schemata as intermediate between sketches and fully characterized mechanisms.

I think there is a better interpretation of MDC's schemata, which is more consistent with how Darden and Machamer describe schemata in other places. The contrast between sketch and schema in terms of our state of knowledge can be made "for epistemic purposes," but the inclusion of that preamble hints that scientists might also have other purposes. Darden makes the epistemic distinction between models of mechanisms at different stages of development in terms of a box metaphor: "The goal in mechanism discovery is to transform black boxes (components and their functions unknown) to gray boxes (component functions specified) to glass boxes (components supported by good evidence), to use Hanson's (1963) metaphor. A schema consists of glass boxes; one can look inside and see all the relevant parts" (Darden 2008). But Darden also describes another, arguably more important, job for schemata to do: "Schema instantiation provides an abstract type of mechanism that may be specified to apply to a particular case. Types of mechanisms may be depicted in abstract mechanism schemata; instantiation is the process of making a schema less abstract and applicable to a particular case" (Darden 2002 S355). This is one of three reasoning strategies she describes for discovering mechanisms.

Darden describes how analogous theories can be grouped together and an abstract schema can be constructed that covers them all. Wave phenomena and natural selection are two examples she mentions. She does not go into much detail here, but presumably the reason for describing types of mechanisms is that they have shared features, which can be explained in the same way. What we know about one kind of wave might turn out to be true of waves in general. This allows us to apply what we learn from experiments in water basins to electromagnetic radiation, for example. Furthermore, each new mechanism we discover provides a new way of approaching phenomena for which we as yet have no good explanation. Darden cites the example of how the discovery of reverse transcriptase "opened up a space of possible mechanisms with feedback into DNA from elsewhere," such as proposals for directed mutation, and feedback from the soma to the immune system (Darden 2002 S361).

What I take from this is that once we've described a mechanism adequately in one context, we might represent it abstractly as a schema for various purposes. As a representation of the original mechanism, it is a glass box that we can look through to see the details. This might serve many practical and cognitive purposes such as communicative ones. As a representation of a type of mechanism, it can also serve epistemic, explanatory and theoretical purposes; we can use it to transfer knowledge about one instance of the type to others (or the type in general), or to form a hypothesis for how a phenomenon we don't yet understand might work. In these latter cases, where the schema is used as a template that can be applied to many contexts, the schema no longer acts as a glass box. Depending on how much we know about the new context, we might be able to fill in the new details right away and immediately have an adequate explanation. Much more likely is that we will be unsure of some of the relevant details, or unsure of the hypothesis that the borrowed schema applies to this case. Here what was a schema of the original mechanism becomes a sketch of the new one. We need to do some work to fill in the gaps and evaluate whether the schema-cum-sketch is helpful.

13.3 Piaget's Schemata

Piaget's schemata work in very similar ways. His theory of how children develop intelligence goes roughly as follows: Intelligence is the result of a complex interaction between the child's environment and their reasoning abilities (which are partly biologically determined, partly a result of this ongoing interaction). Through a series of experiments that Piaget performed on his children (with the help of their nurse), he observed how sucking and rooting instincts gradually develop into coordinated nipple-finding skills, how undirected looking develops into visual acuity, and so on.

The sucking schema, in Piaget's simplest example, begins as a reflex to suck whenever something touches the insides of the lips. The infant will try sucking blankets, fingers, or other body parts if it comes into contact with those. The schema gradually becomes more complex with experience. New things the baby encounters, such as its own thumb, might get added to the schema, and its movements become more coordinated such that it can automatically turn its head in the direction of the nipple, and move its thumb to its mouth. This requires a process Piaget calls *assimilation*, which involves cumulative repetition, generalization (incorporation of new objects), and finally motor recognition (Piaget 1952, 36–37). A second process called *accommodation* can alter the schema to better account for experience. When the schema does not lead to satisfactory results, for instance when the infant sucks something it doesn't like, accommodation will ensure that the infant won't try sucking the offending object again. The coordination of two schemata, such as is required for recognizing the nipple by sight, also involves accommodation. These processes of assimilation and accommodation working together in equilibrium are what Piaget calls *adaptation*.

In Peter's analogy, "understanding the world through Piaget's schemata is like understanding the world through or by mechanisms" (personal communication). We can think of the mind of the child as being like the state of scientific knowledge, and the schemata through which the child makes sense of their world as akin to the mechanism schemata through which scientists make sense of nature. Both start out by fumbling around taking in information indiscriminately, gradually collecting experiences, grouping them together, and drawing associations between them, which coalesce into hypotheses about how the world works, or schemata for understanding it. Both continue gathering information, taking note of exceptions, and making alterations to the schemata when surprises crop up. Paraphrasing Peter, we can think of some schemata (causal schemata in particular) as ideas of mechanisms. He says, "Sometimes the schema you need to understand something is the mechanism by which it works... And since schemata are always modifiable with new input (accommodation) they are always in a sense preliminary and in need of development" (ibid).

This is quite a bit different than the popular interpretation of mechanism schemata as glass box descriptions of well understood mechanisms, or later stages in the development of a sketch. Sketches typically are used to represent the (incomplete) state of our knowledge. They only apply to one phenomenon, and usually only increase in epistemic support. Schemata, on the other hand, are not primarily employed as representations, and part of their purpose is to be widely applicable. An emphasis I would add is that the primary purpose of schemata (both Piagetian and mechanist) is that they can be used to make sense of experience. Piagetian schemata are supposed to help us understand new experiences, so applicability to various cases is their very purpose. The more we can understand with a given schema, the better a schema it is.[2]

Schemata are formed based on what we have already experienced, so in a sense they represent that past experience, but the purpose of them is to help us understand what we have yet to experience (or what we were not yet able to understand when we did experience it). Going back to Darden's point, we develop mechanism schemata (at least in part) in order to be able to instantiate them in novel ways. A mechanism schema might start out as an abstract description of a particular phenomenon, but its main purpose is to help us connect that phenomenon to other things. These connections may be to phenomena we already understand, or to new phenomena that we don't yet know how to understand. They are more like working hypotheses, with the emphasis on *working*. Schemata may fall anywhere on the spectrum of epistemic support from complete guesses to well-supported theories, and the same

[2] I do not mean to say that generality is the only or even the main characteristic that we look for in schemata. A schema that can make sense of a previously mysterious experience is also valuable, even if that schema is only narrowly applicable. Bogen (2005) defends a similar point.

schema might have a very different epistemic status when applied to different contexts.

Examples of this use of mechanism schemata are as common as analogical reasoning in science, which is to say utterly ubiquitous. An example I like is Hinton's 'wake-sleep' algorithm. The 'wake-sleep' algorithm is a method for unsupervised training in neural networks, i.e., a way for a vaguely brain-like artificial intelligence system to teach itself how to improve its performance without the need for a teacher or a set of training data to explicitly tell it what it did wrong or right. One way in which a schema from elsewhere is used in this algorithm is apparent from the name; the training method was inspired by psychological theories about the function of dreaming. The algorithm cycles between 'wake' and 'sleep' phases. In the 'sleep' phase, the network generates stochastic 'fantasy' vectors at the output layer. "These fantasies provide an unbiased sample of the network's generative model of the world" (Hinton et al. 1995). In other words, the fantasies represent the network's encoding of its inputs up to that point in training. Based on these fantasies, the network's parameter settings are adjusted such as to maximize the probability of recreating the input that could have produced that fantasy as output. Metaphorically speaking, the network tries to make sense of its thoughts by changing the very machinery that gives rise to them in such a way as to make those thoughts unsurprising.[3] Furthermore, the method Hinton uses for adjusting the parameter settings is a schema borrowed from physics. The method is "precisely analogous to the way in which the energies of the alternative states of a physical system are combined to yield the Helmholtz free energy of the system" (Hinton et al. 1995).

Hopefully the idea is clear despite this being a rather technical example. A hypothesis about how dreaming works, and a bit of math developed in an unrelated branch of physics were redeployed here to do different jobs than they were originally developed to do. Instead of starting from scratch and building a neural network training method from the ground up, Hinton borrows ideas that he knows already work to solve new problems. To put it bluntly, the kind of science that gets published in *Science* creatively adapts successful ideas from other fields, abstracts away the details, then re-applies them to new cases. Mechanism schemata are the abstract models of mechanisms that can be so deployed.

13.4 MDC's Anti-abstraction Reputation

Darden and Machamer's view(s) of schemata highlight these ways in which abstraction is useful in mechanistic explanation. Curiously, several recent papers which also argue for the neglected role of abstraction in mechanistic explanation cast MDC as their opponents.

Kuorikoski argues that there are two distinct concepts of mechanism: "First there is the concept of mechanism as a componential causal system, which is accompa-

[3] Incidentally, this is not unlike Piaget's account of the development of intelligence.

nied with the heuristics of decomposition and localization. Second, there is the concept of mechanism as an abstract form of interaction, accompanied by the strategy of abstraction and simple models" (Kuorikoski 2009). In cases from economics, he shows that it is relational properties between entities rather than properties of their parts that do the explaining. He goes on to say that, "Since the form of interaction is not in itself dependent on the way the causally relevant properties of the component parts are constituted..., the same sample models and hence 'the same' mechanism schemata can be utilized in many different kinds of contexts or domains" (Kuorikoski 2009, 152). The problem Kuorikoski raises is not just that we seem to want to use the term 'mechanism' for both token causal chains and for abstract types of interactions, but furthermore, that in many branches of science, the factors that seem to do the work in explanations are not the details about the parts, but rather relational properties between these parts. He complains that recent accounts of mechanism are primarily concerned with the componential causal system concept, and calls out MDC as being "especially" guilty in this regard (Kuorikoski 2009, 147).

Levy and Bechtel (2013) also argue for the value of abstraction in explanation, focusing on the Hodgkin-Huxley model. They complain that accounts of mechanistic explanation miss cases where "a model is *deliberately* 'sketchy', i.e. where gaps aren't the product of ignorance or theoretical limitations, but of an intentional strategy" (Levy and Bechtel 2013). This sounds like what Peter had in mind for mechanism schemata; however, MDC are named as a set of authors whose line of thinking requires correction because they tend to "emphasize completeness and specificity" (Levy and Bechtel 2013). The cases Levy and Bechtel cite as benefiting from more abstract explanations are ones where lower-level entities are treated as collections or aggregates, such that the individual details of each entity are less important to the explanation than the properties of the collective. This is of course just one way in which abstraction from details can be useful. There are many other abstract types that schemata might capture beyond just aggregates, as Darden's examples illustrate.

Since Craver has been the most prolific of the three in expanding on the MDC view of mechanism, and his 2007 comments about schemata seem to take a different direction, it is understandable that the role of schemata as abstract mechanisms in the MDC account has largely escaped notice. In several places (Craver 2006; Craver 2007; Kaplan and Craver 2011) Craver even seems to suggest that explanatory adequacy cannot be achieved with abstract models. He claims that "merely subsuming a phenomenon under a set of generalizations or an abstract model" does not explain the phenomenon (Craver 2006). He warns against models that provide an equation that describes the phenomenon, but without specifying what the underlying mechanisms are that make the equation a good fit. He criticizes how-possibly models, or "loosely constrained conjectures about the mechanism that produces the explanandum phenomenon" (Craver 2006), and argues that Hodgkin and Huxley's equations describing ion flow during the action potential were not truly explanatory until correspondences between the equation's parameters and entities acting in cell membranes were confirmed (Craver 2008).

I take it that these claims are concerned not so much with whether abstract descriptions of mechanisms play any explanatory role at all, as with whether abstract descriptions are sufficient to explain. Craver's point is perhaps not that abstract models have no explanatory value, but rather that for descriptions of any kind to be explanatory they must refer to what's really making the explanandum happen. Craver is often interpreted as arguing for a 'more details better' account of mechanistic explanation, according to which mechanisms must include all the gory details that could under any circumstances be relevant to their operation in order to be explanatory.[4] I take Craver's point about detail to be a (not entirely convincing) account of how to draw boundaries around mechanisms, not about how much detail is required in any particular mechanistic explanation. In epistemic accounts of explanation it is all too easy to find a role for abstraction: abstract descriptions, metaphors, and analogies are all excellent ways of communicating information about mechanisms, regardless of whether anything in the world literally corresponds to them. MDC's explanatory concerns are focused on the mechanisms themselves, however, and their productive continuity. Craver later recasts productive continuity in terms of Woodward's ((2003)) interventionist account of causation. The MDC account is not anti-abstraction; it just requires that abstract models refer to the actual difference-makers in order to be considered explanatory.

13.5 Abstract Schemata and Explanation

What remains is to show how abstract mechanism schemata can act as explanations, which for MDC implies that they refer to the actual difference-makers. It should be uncontroversial that at least sometimes we need the gory details to explain how a mechanism works. To see why we needn't always go into all the details, and how details and abstraction might be combined, it is helpful to compare the 'more details better' argument attributed to Craver to Kim's argument against non-reductive physicalism (Kim 2005).

In Kim's argument, the possibility that there might be autonomous mental causation acting at a higher level is rejected, because the lower-level physical events they supervene on also seem to be causes of subsequent events, and causes shouldn't be overdetermined (by assumption). Just as Kim chooses to have mental causes be reducible to physical ones in order to avoid his dilemma, the 'more details better' interpretation of Craver locates the real difference-makers in the gory details. Where the comparison breaks down is that an essential assumption in Kim's argument is causal closure of the physical domain. In the MDC account, both higher and lower-level mechanisms are physical, so causal closure does not block the possibility that a lower-level event could be caused by a higher-level one. The physical cause of the lower-level event has to also be physical, according to causal closure,

[4] See Craver and Kaplan (in preparation) for Craver's most recent account on this.

but needn't be at the same mechanistic level. Causal relationships can cross-cut mechanistic levels.

Machamer and Sullivan (2001) question the possibility of unabmiguously identifying mechanistic levels because of the ubiquity of cross-level relationships. There are cases where causal relationships cross levels, and cases where particular entities can appear in multiple levels alongside other entities of different scales. As I argued in Stinson (2016), there is little reason to believe that mechanisms must be organized into neat hierarchies following mereological relationships. Instead of talking about levels of mechanisms, I'll talk in terms of a continuum from fine-grained detail to abstraction.[5]

This rejection of an analogue to Kim's causal closure argument, and the recognition that mechanistic levels may cross-cut each other both in terms of composition and causal relationships suggests another way to solve the puzzle. Certainly some of the causes at play in a mechanism might be fine-grained ones. These would typically be responsible for small effects that would only be relevant in some explanatory contexts. But other causes at play in a mechanism might be coarser-grained ones. These would typically be responsible for the major effects and would be relevant in most explanatory contexts. (Of course small details sometimes have large effects, and vice versa.) One of the reasons why mechanism schemata are so useful is that they pick out the major difference-makers, so focus attention on the features of the mechanism that are most important.

As described above, schemata start out as glass box descriptions. These are abstract representations of mechanisms that we already understand. Some details are removed in order to show more clearly how the mechanism works. Schemata are often assumed to have the potential to be applied more widely. They can act as hypotheses for how unknown mechanisms may work in what Darden calls schema instantiation. Here they are turned into sketches. These sketches need to be filled in and empirically justified in order to gain the status of explanations. If in the new context the schema-cum-sketch turns out not to correspond to the underlying mechanisms, then it is not an explanation. This is the sort of case Craver warns against. However, when the sketch does prove to have been successful, the original schema explains the new phenomenon too. (It may also happen that the original schema gets altered to fit the new case, which may result in the old explanation being adjusted too. This can be understood in terms of Piaget's assimilation and accommodation.) When a more abstract model can serve to unify the explanations of several particulars, this is usually thought of as an explanatory virtue. Although I wouldn't endorse the view that unification is *the* goal of explanation, it is one among many explanatory virtues. Of course an abstract schema will rarely be a complete explanation of either the original or the new phenomenon. Some of the details of each case might also be required in a complete or even adequate explanation. Most adequate explanations

[5]The continuum of granularity that Weiskopf (2011) describes refers to "size of the chunks into which one decomposes a mechanism." My continuum of abstraction is closely related, but refers to how much detail we include in our descriptions, regardless of the size of the components. Both small and large things can be described in more or less detail.

are probably combinations of multiple models at various levels of abstraction. Different combinations of these multiple models would be called for in different explanatory contexts. Nevertheless, the abstract schema is an explanatory model in virtue of being part of the explanation.

To give a very simple example, a bag filled with sand used as a draught stopper at the base of a door does its job of stopping draughts in virtue of it being heavy, dense, and heat resistant enough not to let cold winds blow in. Each grain of sand in it does part of the job, but it's the fact that it's a lot of them bound together in a bag that matters most to it serving its purpose. This is a case of an aggregate, like the ones Levy and Bechtel discuss. In some other contexts, the finer-grained properties of the sand might matter. If we wanted to know how the sandbag's draught stopping power changes after being hit by lightning (an exceptional case of the type Craver might worry about), it would be relevant to know what proportion of silicon dioxide and calcium carbonate the sandbag contains. If the particular shape of the glass created by the lightning strike also affected the sandbag's draught stopping properties, it might be necessary to know the distribution of grains of particular chemical compositions in different parts of the sandbag. In each case we have the same bag of sand, but what matters about that bag of sand shifts depending on the explanatory context.

Regardless of how much detail we might need, a complete description that includes all the causally-relevant aspects of the mechanism needn't be a collection of just fine-grained entities and activities like an exhaustive description of every grain of sand. It would instead be a collection of models at various levels of grain and abstraction. It might include things like the mass, shape, and wind permeability of the whole bag, and the distributions of the different particle types in the bag, in addition to the locations of individual grains. We can leave the finer-grained details out of the explanation in cases where the locations of individual grains don't make a difference. But we also want to maintain an explanation that corresponds to the actual difference-makers. In these cases, the explanation might just include the mass and shape, or the distributions of grain types. Abstract explanations like these still refer to the actual difference-makers.

One minor quibble is that it is not clear here whether what we call 'the mechanism' is what is picked out by the complete description regardless of the explanatory context, or whether what constitutes the mechanism shifts as the context changes to only include the components that are causally relevant in that context. In one sense these pick out the same thing in the world. The sandbag as a massive object of a particular shape doesn't leave out any of the grains. We can use schemata to pick out the relevant aspects of the mechanism and abstract away the details. This abstracting away doesn't lose details the way one might lose grains of sand if they were to spill out a hole in the sandbag. Schemata still capture the entire mechanism, just under a coarser-grained description.

Schemata aren't the loosely constrained conjectures about mechanisms that Craver warns against. They are not just less sketchy sketches. They have a different epistemic, explanatory, and theoretical status. One might know in great detail about a mechanism and the related components and sub-mechanisms that surround it, but nevertheless have good reason to think that in some contexts what makes a difference is just a coarse-grained aspect of that cluster of components. It might only be a mechanism's information-processing properties that make a difference to what it does, and not how those are implemented. It might only be the fact that it obeys the Hodgkin-Huxley equations that makes a difference, and not the finer details of exactly what makes it obey those equations. The most important difference-makers are often abstract features of the system, as Kuorikoski and others have pointed out. These abstract features of mechanisms are what schemata describe, and as MDC say, this is what scientists are typically interested in.

Take the somewhat more complex example of a unicycle. Insofar as the unicycle is an instance of a general gyroscope schema, its angular momentum is conserved and it stays upright. It works much the same way as the gyroscopes in helicopter navigation systems and iPhones. Insofar as its wheel is out of true, it wobbles. Although there might be some explanatory contexts where finer details about the wheel are needed, the more general facts are not jettisoned; they are added to. If the ground is smooth, angular momentum might be enough to explain why the unicyclist stays upright. If the ground is bumpy, the fact that the wheel is out of true might also be required for an adequate explanation.

The role that abstract mechanism schemata play in discovery carries over to explanation. What mechanism schemata do is highlight the features of a mechanism that are the most important difference-makers, and show how they work in general. When we abstract the details away from a mechanism to create a schema, it should pick out these important features. If we then instantiate the schema in a new context, the hope is that the features highlighted in the schema work in roughly the same way as the features playing the same role in the original mechanism. If the instantiation is successful, we assimilate the new case into our schema. We might also find out something new about the schema in the process of trying to instantiate it, and alter the schema to accommodate that new knowledge. In this sense schemata are always open to change, as is all of our scientific knowledge.

13.6 Conclusion

I began with a disagreement over how to pluralize 'schema' and a mysterious remark about Piaget. I drew together M, D, and C's various comments on schemata from other texts to get a more accurate picture of their view(s) than the popular reading. By combining Darden's work on schema instantiation in mechanism discovery with Peter's Piagetian inspiration, I constructed a picture of mechanism schemata as abstract templates that are formed by abstracting away details of experience, and that are used to help understand new experiences. Craver's focus on details and insistence that explanations reflect the actual difference-makers are often taken as precluding any explanatory role for abstract models or generalizations. I showed how abstract models like mechanisms schemata can reflect the actual difference-makers, and be combined harmoniously with more detailed models. The role of schemata in mechanism discovery carries over to explanation. Because of how they are constructed through a process of ongoing assimilation and accommodation, schemata are uniquely placed to serve as hypothesized explanations. Once they receive empirical support in new contexts, mechanism schemata become genuine, albeit partial, explanations.

References

Bogen, Jim. 2005. Regularities and Causality: Generalizations and Causal Explanations. *Studies in History and Philosophy of Biological and Biomedical Sciences* 36(2): 397–420.

Craver, Carl F. 2006. When Mechanistic Models Explain. *Synthese* 153(3): 355–376.

———. 2007. *Explaining the Brain: Mechanisms and the Mosaic Unity of Neuroscience*. Oxford: Oxford University Press.

———. 2008. Physical Law and Mechanistic Explanation in the Hodgkin and Huxley Model of the Action Potential. *Philosophy of Science* 75(5): 1022–1033.

Craver, Carl F. & Kaplan, David Michael. in preparation. Are More Details Better? On the Norms of Completeness for Mechanistic Explanations.

Darden, Lindley. 2002. Strategies for Discovering Mechanisms: Schema Instantiation, Modular Subassembly, Forward/Backward Chaining. *Philosophy of Science* 69(3): S354–S365.

———. 2008. Thinking Again About Biological Mechanisms. *Philosophy of Science* 75(5): 958–969.

Hinton, Geoffrey E., Peter Dayan, Brendan J. Frey, and Radford M. Neal. 1995. The 'Wake-Sleep' Algorithm for Unsupervised Neural Networks. *Science* 268 (5214). American Association for the Advancement of Science, 1158–1161.

Kaplan, David Michael, and Carl F. Craver. 2011. The Explanatory Force of Dynamical and Mathematical Models in Neuroscience: A Mechanistic Perspective. *Philosophy of Science* 78(4): 601–627.

Kim, Jaegwon. 2005. *Physicalism, or Something Near Enough*. Princeton: Princeton University Press.

Kuorikoski, Jaakko. 2009. Two Concepts of Mechanism: Componential Causal System and Abstract Form of Interaction. *International Studies in the Philosophy of Science* 23(2): 143–160.

Levy, Arnon, and William Bechtel. 2013. Abstraction and the Organization of Mechanisms. *Philosophy of Science* 80 (2). The University of Chicago Press on Behalf of the Philosophy of Science Association, pp. 241–261.

Machamer, Peter K., and Jacqueline A. Sullivan. 2001. Leveling Reductionism. http://philsci-archive.pitt.edu/id/eprint/386.

Machamer, Peter, Lindley Darden, and Carl F. Craver. 2000. Thinking About Mechanisms. *Philosophy of Science* 67(1): 1–25.

Piaget, Jean. 1952. *The Origins of Intelligence in Children*. New York: International Universities Press.

Stinson, Catherine. 2016. Mechanisms in Psychology: Ripping Nature at Its Seams. *Synthese* 193(5): 1585–1614.

Weiskopf, Daniel A. 2011. Models and Mechanisms in Psychological Explanation. *Synthese* 183(3): 313–338.

Woodward, James. 2003. *Making Things Happen: A Theory of Causal Explanation*. Oxford/New York: Oxford University Press.

Chapter 14
MOCing Framework for Local Reduction

Tom Seppalainen

Abstract In sensory neuroscience, the neural and perceptual levels of investigation are commonly related through a reductive research strategy based in psycho-neural isomorphisms. Davida Teller's "linking propositions" are a particularly vivid illustration of this epistemology in the context of vision science. For Teller, linking propositions guide the core epistemological practices of vision science by expressing the criteria for acceptable explanations of perceptual phenomena by neural processes and by articulating heuristics for discovering neural properties on grounds of perceptual ones, and vice versa. Furthermore, linking propositions create a uniform discourse in which ontological identity statements can be formulated on the basis of structural terms applicable to both the neurophysiological and perceptual domains.

I argue that the epistemic practices of current color vision science are better understood through Machamer-style mechanisms than isomorphism-based linking propositions. For Machamer, activity types of mechanisms are central for both explanation and the discovery of entities and their properties. The activity type of color cancellation is fundamental to the opponent process paradigm of color vision precisely in this manner. The descriptive inadequacy of linking propositions for clarifying vision scientific research is particularly damaging because activity types have a unifying role for the structural facts of color perception. As a form of local reductionism, Teller's proffered epistemology aims to create a uniform domain for neuroscience and perceptual psychology but on the basis of a non-unified, disparate set of structural identity statements. By bringing attention to the unifying role of activity types, Machamer mechanisms allow also for critical epistemological explorations of ontological reductions on grounds of the fit between activity types of higher-level and lower-level sciences. They offer a Machamer-ontological-critical or *MOCing* framework, for short, against local reductionism.

T. Seppalainen (✉)
Portland State University, Portland, OR 97201, USA
e-mail: seppalt@pdx.edu

M.P. Adams et al. (eds.), *Eppur si muove: Doing History and Philosophy of Science with Peter Machamer*, The Western Ontario Series in Philosophy of Science 81, DOI 10.1007/978-3-319-52768-0_14

14.1 Introduction

Are there specific concepts for discovery and explanation that vision scientists use to clarify the manner in which they go about creating a unified domain from the neural and perceptual levels of scientific investigation? And if there are, do these concepts facilitate critical analysis of the scientific practices aimed at discovering the neural underpinnings of visual phenomena? In this paper, I will explore these two interrelated questions, one descriptive and the other normative, through Davida Teller's ideas on the nature and role of "linking propositions" in vision science.[1] Linking propositions express structural identities and resemblances between mind and brain states and form the core of a specific reductive research strategy based in psychoneural isomorphisms. They create a uniform discourse for the neurophysiological and perceptual domains on the basis of a conceptual scheme of relational or structural notions, one in which identity statements can be formulated. They also guide the core epistemological practices of vision science: linking propositions express the criteria for acceptable explanations of perceptual phenomena by neural processes and articulate heuristics for discovering relevant neural properties on grounds of perceptual ones, and vice versa.

Teller's isomorphism model of the epistemological practices of vision science is an example of local reductionism. Local reductions offer identity statements without a requirement of a unifying theoretical framework capable of explaining a multitude of phenomena. Teller's model exemplifies local reductionism in presenting an open-ended list of linking propositions consisting of various relational characteristics such as identity, mutual exclusiveness, simplicity, and spatial analogy.

In this paper, I will argue against Teller's model of vision scientific research on two interconnected grounds. I will undermine its descriptive adequacy by arguing that her linking propositions do not clarify the research heuristics that govern the experimental practices in the opponent process theory of color vision, her prime example of the use of all linking propositions. Instead of a set of linking propositions, the integration of perceptual and physiological domains of color occurs through a unifying concept of color processing, cancellation. I will show the central role of cancellation in both explanation and the experimental measurement practices designed to discover and identify entities and properties of the mechanism of color vision. On the basis of the case study, I show that Machamer's mechanism-based model for scientific epistemology is fitting for describing both the nature of explanation and the investigative aims and tools of vision science.[2] The model fits because it confers activity types a central role in scientific practices.

[1] Teller D. Y. (1980), "Locus questions in visual science", in C. S. Harris (ed.), *Visual coding and adaptability*. Hillsdale, New Jersey: Lawrence Erlbaum, 151–176; Teller, D. Y. (1984), "Linking Propositions." *Vision Research*, 24, 10, pp. 1233–46; Teller D. Y. (1990), "The domain of visual science," in L. Spillman and J. S. Werner (eds.), *Visual perception: The neurophysiological foundations*. New York: Academic Press, 11–21.

[2] See, e.g., Machamer, 2003. (Machamer, Peter. 2003. "Activities and Causation: The Metaphysics and Epistemology of Mechanisms." *International Studies in the Philosophy of Science*, Vol 18, 27–39).

Teller's version of local reduction is also insufficient for normative epistemological purposes. Linking propositions aim to create a uniform domain for neuroscience and perceptual psychology but endeavor to accomplish that on the basis of an ad hoc set of structural identity statements. This epistemology violates standard criteria of evaluation of scientific hypotheses not to mention reductionist ones. I will argue that Machamer's mechanistic model has the conceptual resources to solve these problems. A Machamer-mechanism explains when its activity types integrate visual phenomena including relational ones and explains reductively when the postulated activity types are ontologically feasible, possible from the standpoint of the activity types of mechanisms of lower-level sciences. Due to its descriptive potential and normative nature, this mechanistic conceptualization of vision science generalizes to a *MOCing* or "Machamer-ontological-critical" framework for local reductionist efforts.

14.2 Local Reductionism in Vision Science: Teller's Linking Propositions

According to Teller, explanation in vision science consists of a sequence of "mappings." Most of these describe links in the "causal chain" of visual processing that extend from stimuli, through the retinal image and receptors, to physiological states at different stages of neural processing.[3] In addition, explanation requires "linking propositions" that concern the "final mapping" between visual experiences (or behavior) and the neural substrate.[4] The role of linking propositions is to bring neuroscience and perceptual psychology into one "single viable scientific domain" or, alternatively, to create "a single realm of discourse."[5] They accomplish this by specifying particular psycho-physiological mapping principles according to which some broadly speaking relational characteristic of neural processing implies an identical or similar relational characteristic for perceptual phenomena.[6] Teller identifies five general "families" of linking propositions used in vision science: *identity, similarity, mutual exclusivity, simplicity, and analogy*. In this section, I will, first, characterize these families in a manner that clarifies Teller's linking propositions and, second, explore their epistemological use and ontology in order to elucidate why linking propositions exemplify local reductionism.

According to the base form of the identity family of linking propositions, identical neural states entail identical perceptual states or sensations. The converse identity proposition is the most important principle of this family in vision research. According to it, identical or indistinguishable sensations entail indistinguishable

[3] See, e.g., 1980, 152 and 1984, 1235.

[4] 1984, ibid.

[5] Ibid, 1233, and 1990, 20.

[6] 1984, 1235.

physiological states. As an example of its successful use, Teller mentions metamers, lights of different wavelength composition that, despite their physical distinguishability, appear indistinguishable in color or hue.[7] Metamers led vision scientists to postulate only a "few kinds of receptors" into the visual system.[8] Similarity comprises another family of linking propositions. According to the converse similarity formulation, similar sensations imply similar states of the nervous system. Inferring similar neural codes for long and short wavelength lights of the visible spectrum on grounds of their similarly reddish appearance illustrates its use.[9] Mutual exclusivity covers another family of linking propositions.[10] The converse mutual exclusivity proposition is used in the opponent process theory of color vision to establish two neural color channels that can take on either of two mutually exclusive, opponent states on grounds of the mutually exclusive hue sensation pairs, red-green and blue-yellow.[11] According to the fourth, simplicity family of linking propositions, simple physiological states link with simple perceptual states. Its use is illustrated by the explanation of unitary hue sensations through activation in only one of the neural opponent color channels.[12] Simple sensations imply a particularly simple neural code.

In addition to what she calls "rigorous" linking propositions, Teller identifies a fifth, open-ended family of analogy linking propositions.[13] It concerns "pictorial or graphical analogies" between data sets.[14] The use of analogies occurs through heuristics that make perceptual and physiological phenomena "look alike."[15] The resulting concept of explanation is premised on a "looks like" criterion: if neural and perceptual phenomena look alike then the former explains the latter.

The above brief illustrations of the use of linking propositions indicate they have significant epistemological and methodological roles in vision science. Further analysis of their use in explanation and discovery together with an interpretation of their ontology demonstrates they illustrate reductionism in a specific, "local" form.

Identity, similarity, mutual exclusivity, simplicity, and graphical analogy all concern relational characteristics that apply at both the perceptual and neural level. Linking propositions map the perceptual to the neural domain on such grounds. The explanation of the former with the latter occurs because of the relational identities and/or resemblances across the two domains. Formally, the ensuing concept of explanation fits best a deductivist model and Teller expresses all linking propositions through logical entailments. Furthermore, linking propositions display the logic of identity. All other neural processes than the "immediate substrate" that link-

---

[7] 1984,1237.

[8] Ibid.

[9] Ibid, 1237–1238.

[10] Ibid, 1239.

[11] Ibid.

[12] Ibid, 1240.

[13] Ibid.

[14] 1990, 13.

[15] Ibid. See also 1980, 152–155.

ing propositions refer to are "not necessary" for a specific perceptual state.[16] Furthermore, Teller explains that the perceptual state would occur upon stimulation of the neurons referred to in the final mapping even in the absence of stimulation of other neural stages.[17] The immediate substrate is sufficient for the perceptual state. Thus, linking propositions establish psychoneural reductions with identity statements.

Linking propositions form the methodological core of a reductionist research strategy or heuristic. Traditionally, psycho-neural isomorphism assumptions led to reductionist theorizing about neural processes but exclusively on grounds of perceptual phenomena. They comprised a one-way tool of inference used to speculate about otherwise inaccessible neural properties on the basis of facts of perceptual psychology. Teller's view of reductionist research efforts governed by linking propositions is different. First, the availability of detailed and direct studies of the response properties of neurons has not lessened the importance of isomorphism-based linking propositions.[18] Second, linking propositions license two-way inferences, from perceptual to the neural domain, and vice versa. This is not a mere logical possibility or consequence of the identity statements between neural and perceptual domains. Teller maintains that the relational characteristics invoked in rigorous linking propositions lend themselves to *intra*-disciplinary identification: one has "only to make judgments about the properties – identity, similarity, etc. – of physiological states with respect to each other and perceptual states with respect to each other."[19] This idea about the nature of the reductive heuristics complements the contrast between classic and current use of isomorphisms. Classic, speculative one-way inference could not involve "judgments" of neural characteristics on physiological grounds since no such knowledge existed. If the relevant neural characteristics could not be "judged" within neuroscience, the use of linking propositions in today's vision science would threaten to remain similarly speculative to the classic ones.

Regardless of how the two-way inferences licensed by linking propositions are carried out, their presence creates a research strategy. Each of the four rigorous propositions is applicable to reasoning about properties at the perceptual level on the basis of research into neural processing. This includes the derivation of predictions about visual phenomena. In short, Teller's linking propositions form a genuine reductive research strategy because the reducing level(s) of neuroscience can lead efforts in the study of visual perception.

Linking propositions disclose a reductive ontology far more literal than that associated with identity statements in traditional theory-reductions. For Teller, linking propositions answer a "where" or "locus" question of some perceptual phenomenon.[20] The answer or "immediate substrate of visual perception" is a "bridge

---

[16] Ibid.

[17] Ibid.

[18] 1984,1233.

[19] Ibid, 1240.

[20] See, 1980, 152–3.

locus."[21] It is a distinct site or stage of neural processing instead of a chain or multi-level, distributed causal process. Corresponding to this literal ontology, traditional "bridge principles" or "bridge laws" are replaced by a "bridge locus," a concept relative to which distinctions concerning the epistemological and ontological aspects of reductionism start to fade away.

The "local" nature of Teller's reductionist vision science can now be described in its main dimensions. Linking propositions concern the neural loci of particular perceptual phenomena. These are typically derived in highly specific experimental set-ups. In other words, the *explananda* themselves are "local" visual phenomena. The same feature applies to the *explanantia*. A bridge locus can occur at any brain location or even beyond that, in receptoral stages of processing. As a consequence, the explanatory and discovery efforts are not constrained by any general, unifying ideas about the, for example, central location of (visual) consciousness. Furthermore, different perceptual phenomena can be assigned neural loci on grounds of different linking propositions depending on which relational characteristic is seen to apply to them. The reductionist research strategy can occur in a piecemeal fashion. Teller draws attention to the fact that her five linking propositions might not be enough for characterizing practices of vision science.[22] From both ontological and epistemological perspectives, the central feature of Teller's reductionism is the absence of unifying features. This state of affairs is a departure from original reductionism that served both ontological (e.g., Putnam and Oppenheim) and epistemological (e.g., Nagel) unification. To indicate this difference, I will use "local reduction" and its cognates to characterize Teller's model.

14.3 Analogies and Reductive Research Strategies

Analogies between neural and perceptual phenomena are central to Teller's reductionist research heuristics for vision science and for two interconnected reasons. The first concerns their epistemological status. Analogies, for Teller, are hypotheses and, hence, subject to the kinds of tests "other scientific propositions" are.[23] The rigorous linking propositions, in turn, are "axioms" or the "premises" of visual science.[24] She also refers to them as "analytic truths" and surmises they are "probably not subject to empirical confirmation or disconfirmation."[25]

---

[21] Ibid., and 1984, 1235.

[22] 1984, 1241.

[23] Ibid, 1240–1241; see also 1980, 156–157.

[24] 1990, 15. See also 1984, 1235.

[25] 1990, ibid. Teller's reductionism differs from recent purportedly descriptively accurate forms of scientific reductionism such as Bechtel and McCauley's heuristic identity theory (McCauley, R. N. and Bechtel, W. 2001. "Explanatory Pluralism and Heuristic Identity Theory," *Theory & Psychology*, Vol 11, 736–760) and the type of mind-brain reductionism that, contrary to the original Smartian intentions, construes identity statements as regular empirical hypotheses (see, e.g.,

The second reason concerns the use of linking propositions. As described in the previous section, Teller maintains that intra-disciplinary conceptual resources are adequate for using rigorous linking propositions. For her, neuroscientists, for example, are able to identify the relevant relational characteristics in the rigorous linking propositions, properties such as mutual exclusivity and simplicity, by, presumably, comparing physiological states with respect to each other. The identification of analogies between the perceptual and neural domains, in turn, requires comparing neurophysiological properties to perceptual ones.[26] The use of the reductive heuristic that results in concepts that refer to both neural and perceptual domains – even if only analogously – is an interdisciplinary effort.

In light of their central role, Teller's analysis of analogical reasoning in vision science lacks detail and leaves the heuristics obscure. According to her, perceptual or psychophysical and neurophysiological phenomena are made to look alike by creating "pictorial or graphical analogies."[27] The heuristic concerns the manipulation of data representations that, in turn, is clarified only in two dimensions: plotting data sets into "meaningfully similar axes" and in a manner that creates "similar shapes" for graphs.[28] How such similarities in data representation are attained in general is left untouched.

In this section, I will clarify Teller's analogy proposition through an analysis of the example case she mentions, the analogy between opponent colors and opponent neural cells.[29] I will show it is based in the rules of measurement used in psychophysical and neurophysiological experiments on color. Reasoning with similarly shaped data representations rests on isomorphic measurement systems. The discussion is intended to remedy a descriptive deficiency in Teller's analysis. In addition, it reveals the real target of the analogy and how it should be conceptualized. The target is represented by the mathematical relationship of subtraction and because this operation structures measurement practices in both psychophysics and neuroscience, its referent is (analogically understood) cancellation activity. I will end the section by illustrating the central role of the analogy of cancellation for the research heuristic comprised of some of Teller's rigorous linking propositions. This illustration undermines Teller's claims about the reductive heuristic in both its use and epistemological status. These interrelated, descriptive and normative epistemological departures from Teller's local reductionism suggest a need for an alternative conceptualization of the epistemological practices of vision science.

How can data from neuroscience and perceptual psychology be compared? Much of the data on perceptual color phenomena concern sensations: basic, primary or unitary ones, their relations, and both to each other and stimuli. The neurophysiological experiments relevant for Teller's linking analogies, in turn, offer data on

Churchland, P. M. 2005. "Chimerical Colors: Some Phenomenological Predictions from Cognitive Neuroscience," *Philosophical Psychology,* 2005, Vol. 18, 527–560).

[26] 1984, 1240.

[27] 1990, 13.

[28] 1984, 1240. See also 1990, 13.

[29] 1984, ibid.

firing rates of neurons and their relationship to stimuli. Although color sensations and neurons' firing are radically different things, experiments on perceptual and physiological phenomena both involve measurement. From the perspective of the empiricist "representationalist" theory of measurement, the data in the two respective disciplines can potentially be similar. The core idea of representationalism associated with the psychometrician S. S. Stevens is that measurement is "the assignment of numerals to objects or events according to some rules."[30] The rules of measurement define a numerical system consisting of numbers and their permissible operations such as order (e.g., larger than), addition, and subtraction. The rules define a measurement structure consisting of numbers and relations.

The main questions concerning Teller's analogy proposition can be addressed through measurement representationalism. Data from neuroscientific and psychophysical experiments can be compared in terms of the numerical systems used to measure and, thus, represent data. Furthermore, the representationalist account makes it understandable how the manipulation of spatial representations of data – the axes for and shapes of graphs – is possible. Visual, spatial representations of and in coordinate systems are derived from numerical systems. Similarities pertaining to the former can be based on similarities pertaining to the latter because structural identities between numerical systems underwrite graphical, "surface" similarities in data representations. The actual heuristic behind Teller's analogy proposition concerns measurement practices and potentially isomorphic, reductionist ones. The creation of isomorphic measurement systems for two central experiments in the opponent process paradigm illustrates the use of such a reductive heuristic.

Hurvich and Jameson's color cancellation experiment is seminal to the opponent process paradigm.[31] The experiment is used to measure the strength of each of Hering's four unitary hue sensations – red, green, yellow, and blue – throughout the visible spectrum. This is accomplished by removing all perceivable traces of each of the unitary hues from spectral stimuli by adding opponent color stimuli at each spectral location. For example, to a stimulus of 590 nm that appears reddish-yellow, a green stimulus is added and adjusted in intensity until the subject reports the disappearance of red and blue is added and adjusted in intensity to remove yellow. The strength of the red hue sensation at 590 nm is measured by the intensity of its opponent hue green needed to remove red and the strength of yellow is measured by the intensity of the blue needed to remove yellow.

Data from the cancellation experiment is represented graphically in two chromatic response curves, red-green and blue-yellow. These two "opponent chromatic response curves" span the whole visible spectrum and both have a positive as well as a negative sign – in keeping with the "cancelling" relationship of the opponent

[30] Stevens, S. S. 1946. "On the Theory of Scales of Measurement," *Science*, 103, 677–680. Reprinted in B. Lieberman (ed.) *Contemporary Problems in Statistics*, New York: Oxford University Press, (1971), 3–8.

[31] See, e.g., Hurvich, L. M. and Jameson, D. 1957. "An Opponent Process Theory of Color Vision," *Psychological Review*, 64, 384–404.

hues.[32] This graphical data representation is based in a measurement structure consisting of numbers attached to paired elements, two unitary hue sensations, and in both, the numbers are related by the arithmetic operation of subtraction.

Neuroscientific experiments in the opponent process theory measure the firing rates of neurons. These are classified further into rates of inhibition and excitation. This distinction is drawn with respect to the rate of response that occurs in the absence of light stimulation, the "spontaneous" discharge rate or "base-rate." During stimulation, spike frequency can be either lower (inhibition) or higher (excitation) than the base-rate. This measurement practice was adopted in color science through the work of R. De Valois and his collaborators.[33] The exact nature of the measurement system requires a precise interpretation of the relationship between inhibition and excitation.

In the opponent process paradigm, inhibition and excitation have different hue functions. The "basic assumption" is that, in each hue cell, excitation is responsible for a unitary hue sensation and inhibition for its opponent hue sensation.[34] The two responses in each of the red-green and blue-yellow hue cells are labeled by the arithmetic signs, "+" and "-." These signs, however, are not mere labels or their quantitative variation only related in terms of order. Instead, the numerical system for firing rates is structured by the relationship of subtraction. This is most readily discernible in the statistical concept of an "average cell." Average cells of both red-green and blue-yellow type subtract rates of inhibition from those of excitation in their respective neural populations to reach average excitation or inhibition rates.[35] Moreover, at a spectral locus where the average inhibition and excitation rates for one of the two (average) cell types is equal, the unitary hue sensation resulting from the (average cell for the) other population of neurons occurs. At that locus, the former cells have canceled out their activation to a level coinciding with the base-rate. The subtractive relation between the two response types represents such cancellation.

---

[32] Ibid.

[33] See, De Valois, 1965. (De Valois, R. I., "Behavioral and Electrophysiological Studies of Primate Vision," in W. D. Neff (ed.) *Contributions to Sensory Physiology Vol. 1*. New York: Academic Press, (1965), 137–178). Earlier, cells were often distinguished into "on-cells" and "off-cells." This is often called "Hartline's principle" [see, e.g., Hartline, H. K. "The response of single optic nerve fibers of the vertebrate eye to the illumination of the retina," *American Journal of Physiology*, 121, (1938), 400–415]. On-cells respond during light stimulation and off-cells after its termination. The "on/off-principle" of classification conceptualizes a cell response exclusively in terms of increases in firing rate and, according to current color science, the ensuing measurement practice denies that inhibition is a real response.

[34] De Valois, 1965, 159. More specifically, the assumption is that "[T]he R + G- cell is signaling red with an increase in activity and green with a decrease in activity, i.e. that the animal will see red when this cell fires rapidly and green when the cell is inhibited."

[35] For more detail, see the analysis given by De Valois et al., 1966, 976 (De Valois, R. I., Abramov, I., and Jacobs, G. H. "Analysis of Response Patterns of LGN cells," *Journal of the Optical Society of America*, 56, (1966), 966–977) of what they call "an isomorphic relationship between the relative activity rates of the various cell types and the hue of a given light."

The numerical system used in neurophysiological measurement underwrites the graphical representations of hue cells. In many, the spontaneous base-rate defines the null-point of the y-axis and, in others, numerical values for inhibition are represented below the spontaneous rate that functions as the virtual null-point.[36] Most importantly, the visual data representations of the perceptual color cancellation experiment and the neurophysiological single cell recording have both "meaningful axes" and can be manipulated into "similar shapes" because of the identical measurement structure used in both.[37] Without the relationship of subtraction in the numerical systems, such comparisons would be impossible. In short, the heuristics articulated by Teller in her analogy proposition are in the opponent paradigm based in the use of isomorphic measurement systems.[38]

The analysis of the two measurement practices suggests in light of the idea that the numerical relations used in measurement represent empirical phenomena that the real analogy between psychophysics and physiology is the referent of the subtraction operation, viz., cancellation. The central role of this analogy can be illustrated through a critical analysis of the purported use of two of Teller's other, rigorous linking propositions, mutual exclusiveness and simplicity.

For Teller, mutual exclusivity linking proposition licenses the inference to mutually exclusive neural responses on grounds of perceptually mutually exclusive hues. Yet this inference on grounds of *intra*-disciplinary conceptual resources does not establish the true meaning of mutual exclusivity. In a purely perceptual sense, mutual exclusivity refers to the fact that red and green, on the one hand, and blue and yellow, on the other, do not occur as perceivable mixtures but, when mixed, appear as de-saturated variants of one of the hues in the pairs. On purely physiological

---

[36] For the two practices, compare, e.g., De Valois 1965, 571, and De Valois et al., 1966, 972.

[37] For the clearest illustration of how neural response functions coincide in axes and shapes of graphs with the psychophysical chromatic response curves, see R.L. De Valois and K.K. De Valois, 1993, 1059 (R.L. De Valois, R.L. and. De Valois, K.K., 1993. "A Multi-stage Color Model." *Vision Research,* 33, 1053–1065).

[38] This analysis provides a critical context for the color subjectivists' (e.g., Hardin and Clark) main premise for the neurophysiological reduction of color sensations or qualia. Consider, for example, Hardin's (1993, 54. [Hardin C. L. (1993), *Color for philosophers: Unweaving the rainbow* (2nd revised edition). Indianapolis: Hackett]) argument for a neurobiological reduction of color sensations: "Some of the chromatically responsive cells show opponent-response patterns that fit psychophysical, inferred chromatic-response patterns beautifully. However, the picture on the individual cell level is not nearly as neat as these selected examples might suggest and statistical analyses have not yet 'nailed down' a quantitative picture to general satisfaction...[T]he opponent scheme may be said to be physiologically confirmed in rough outline but not in detail." The above analysis shows that, first, the two patterns display a visual "fit" because the shape of the neural one was created to match the perceptual one. Second, and as a consequence, the pattern concerning the data of the two graphs does not qualify as a "confirmation" *per se* because of the "theory-laden" nature of the (representation of the) neurophysiological evidence. Finally, Hardin's statistical point and justification by potentially "better" statistical analyses (ibid, xxxii) does not without further argument uplift the concern raised by the theory-laden neurophysiological data. Clark's (see, e.g., Clark A. (1993), *Sensory qualities*. Oxford: Clarendon Press) color reductionism founders on similar grounds. For a systematic critique of color reductionism, see, Seppalainen, T. V. 2001. "Color Subjectivism is not Supported by Color Reductionism," *Philosophica* 68, 901–927.

grounds, in turn, different neural firing rates including rates of excitation and inhibition are all mutually exclusive. No specific response rate can occur simultaneously with any other.[39] But when the operation of subtraction represents mutual exclusivity – as it does in the respective measurement practices in the opponent theory – the canceling meaning of the concept becomes obvious. The use of Teller's rigorous linking proposition of mutual exclusivity reduces to an analogy applicable across the two scientific domains. To put the point in terms of research heuristics, identifying mutually exclusive processes involves an interdisciplinary context of comparison of physiology and psychophysics and through analogical means. Such a comparison is directly represented in the respective sciences' measurement systems by the mathematical operation of subtraction.

The fundamental role of the cancellation analogy can be further illustrated in the use of Teller's simplicity linking proposition. It purportedly mediates the inference from simple or unitary hue sensations to a simple neural counterpart, base-rate firing rate. But the base-rate is no simpler than any other firing rate on neurophysiological grounds alone. Firing rates form a (short) continuum. The base rate becomes a particularly simple state once it is conceived as a null-point or "zero" response. This state, in turn, is the outcome of neural inhibition and excitation canceling each other out. Once the relationship of subtraction is applied to both domains (to represents the analogy-concept of cancellation), the simplicity based linking proposition becomes usable: both unitary hues and neural base rate are simple states since they are outcomes of cancellation (from more complex states). As a result, the use of the simplicity proposition rests on an analogy-concept of cancellation and, thus, involves inter-disciplinary comparisons instead of mere intra-disciplinary heuristics.

Contrary to Teller's image of vision science, the case study shows that an analogy is central to linking the perceptual and neural domains of color vision. Heuristics comprised of linking propositions are unfeasible without an analogy concerning cancellation. Furthermore, this analogy relates the individual linking propositions to one another and to the extent that their status as individual, independent heuristics is undercut. Therefore, the analysis undermines the very use of linking propositions to describe the research heuristics of vision research and precisely because of their local reductionist nature. A central structural analogy evident in the identical measurement practices of experiments in color science replaces Teller's array of independent relational identities and resemblances as a heuristic.

The above analysis also undermines Teller's portrayal of the evidentiary aspects of vision science. According to her, some linking propositions have an axiomatic status and others, the analogy ones, an empirical status. Since the purported rigorous axioms cannot be distinguished from analogies in use, the evidentiary evaluation of vision scientific claims must occur in terms of whatever empirical (and other) criteria are used in the appraisal of falsifiable hypotheses. In conclusion, the

[39] See also Teller's (1984, 1238–1239) distinction between weak and strong readings of the mutual exclusivity proposition.

case study undermines both the descriptive and normative adequacy of Teller's linking propositions in characterizing vision scientific practices.

14.4 Mechanisms and Activity Types

Subtraction is an arithmetic operation that represents cancellation. Cancellation, in turn, is an activity. This suggests that undertakings in vision science are best conceptualized from a perspective to scientific discovery and explanation that emphasizes activity types. Machamer's mechanism-based approach to biology does just that. In this section, I will argue on both descriptive and normative grounds that this conceptualization fits also vision science.

According to the "MDC account," a mechanism consists of both entities and activities.[40] The ontology of mechanisms is explicitly dualistic. Dualism extends also to the epistemology of mechanisms. First, activity types are a way of making phenomena intelligible. Mechanistic forms of intelligibility are part of how mechanisms explain.[41] Second, Machamer prioritizes activities in the identification of entities and their properties.[42] Thus, activities play an important role in the discovery of mechanisms and the evidentiary methods used to confirm hypotheses about mechanisms. This epistemology makes Machamer's mechanisms particularly fitting for describing discovery, confirmation, and explanation in vision science.

In the opponent process paradigm, an activity type makes specific measurements possible. In perceptual psychology, the intensity of color sensations is measured by canceling hues in color mixtures. In neuroscience, the very criterion of a hue cell is based on the canceling relationship between inhibition and excitation. In other words, both entities, such as hue cells, and properties of entities, such as the intensity of hue responses, is discovered through activity types. This heuristic leads to evidentiary methods. The psychophysical and neural measurements offer data in their respective disciplines and, when this data is integrated into the type of phenomena described in Teller's analogy proposition – similarly shaped response curves – these comprise critical forms of evidence for the mechanism of color vision. The current consensus in color science is based on evidence acquired through the exploration of activity types.

The descriptive adequacy of a Machamer-mechanism for vision science extends to the concept of explanation. For Machamer, a mechanism explains by demonstrating how some end state or "termination condition" is produced by the activities of the entities comprising the mechanism. This fits color science. The opponent cells constitute the hue mechanism and explain because, due to their activity types such as cancellation, they are the types of entities that can bring about perceptual end states. For example, the achromatic end state from an initially red color sensation

[40] Machamer, et al. 2000. "Thinking about Mechanisms." *Philosophy of Science*, *67*, 1–25.

[41] Ibid, 21.

[42] See, Machamer, 2003.

upon superimposition of a green light is explained by the increased firing of neurons with inhibitory signals that cancel out the excitatory signals for red. More generally, given a color mechanism with the activity type of cancellation, perceptual relations such as mutual exclusivity and simplicity or unitaryness become seen as termination conditions of the color mechanism. The relational concepts of the perceptual domain refer to "a-temporal" effects of the productive activities of the hue mechanism. Thus, instead of identities or even resemblances between relational concepts of psychophysics and physiology, explanation is achieved by the entities' activity types that bring about robust color phenomena including relational ones.[43]

There are additional benefits than descriptive adequacy to conceptualizing vision science through a Machamer-mechanism. This construct also offers a critical epistemological framework for Teller's local reductionism characterized by an absence of unifying features among linking propositions. The goal of linking propositions is to create a unified scientific domain for perceptual and neural vision sciences through relational identities and resemblances, yet this is supposedly achievable by an ad hoc set of relational concepts. Vision science, for example, can have its own, evolving set of relational identities without these relating to each other or relational ideas of other sciences of perception. This state of affairs is a departure from original epistemological reductionism. It also violates some of the evaluative criteria for other, non-reductionist types of hypotheses. Criteria such as scope, fruitfulness, and avoiding ad hocness indicate that some degree of unification is critical for all scientific hypotheses.

Machamer's epistemology for mechanisms coheres with standard criteria for hypothesis assessment. Activity types lead to relations among phenomena. For example, the activity type of cancellation figures in the explanation of several central relational perceptual color phenomena. The activity type of the color vision mechanism unifies these phenomena and allows the mechanistic hypothesis to satisfy unificationist criteria. And exploring novel relations that should predictably result from hypothesized activity types is how Machamer's mechanisms meet further evidence.

Machamer's conceptualization of mechanisms also suggests further evaluative criteria for particularly reductive hypotheses. Mechanisms with activity types offer not only "*how actually*" but also "*how possibly*" explanations.[44] The articulation of the latter dimension of explanation through mechanistic "forms of intelligibility," activity types that may perhaps even establish (historically relativized) a priori conditions for scientific understanding, suggests that inter-level criteria can play a role in the evaluation of mechanistic explanations. The feasibility of activity types in higher-level mechanisms may become subject to "possibility checks" on the basis

[43] Since space does not permit an investigation of the source of the analogical understanding of cancellation activity in explanation, I suffice with a mere reference to Machamer's views on the need for metaphoric descriptions of activities as an excellent starting point for understanding the case at hand (Machamer, P. 2000. "The Nature of Metaphor and Scientific Description," in F. Hallyn (ed.) *Metaphor and Analogy in the Sciences*. Kluwer: Netherlands, 35–52).

[44] Machamer et al., 2000, 21.

of the activity types of mechanisms of lower-level sciences. Whatever the exact form of such an evaluation, it does prevent explanation from reducing to evidence and, thus, the satisfaction of standard empirical criteria for the evaluation of hypotheses. In other words, Machamer's mechanisms specify an ontic concept of explanation because the ways in which phenomena are made intelligible by activity types are particularly suitable for evaluation on ontologically reductive grounds, on grounds of the activity types of more basic or other relevant sciences. Taken together, the epistemologically and ontologically integrating functions of Machamer's mechanisms that are grounded in activity types offer a critical framework for analyzing the explanatory and investigative aims of local reductionism. Such mechanisms provide a Machamer-ontological-critical or *MOCing* framework, for short, against local reductionism.

14.5 Concluding Remarks

I have argued that the central epistemic practices of vision science are better understood through the discovery of mechanisms and a commitment to mechanistic explanation through activity types than isomorphism-based linking principles. No doubt the *MOCing* framework is relevant for investigating many other areas of sensory and cognitive neuroscience due to the prevalence of isomorphism-based local reductionist practices across them. It vitiates all materialistic, "process Cartesian" solutions to the mind-body problem in which brain structures are in one way of another reflected in phenomenology or simply introspected through consciousness. There is no justification for the concept or property identities resulting from such reductionism according to the ontology of Machamer's mechanisms. Scientific approaches to the mind that satisfy the concepts and conditions of the *MOCing* framework have still plenty to offer for metaphysical questions about the mind, at least once concepts for the activities and movements of the mind are developed through the kinds of inter-disciplinary resources needed for identifying and discovering their mechanical underpinnings – most likely both within and beyond the skull.

Acknowledgement Sincere thanks to Jacqueline Sullivan and Uljana Feest for many helpful comments, suggestions, and criticisms. And thanks to Peter Machamer for his example, insight, and inspiration – both in history and philosophy of science and enology.

Kolibri 360 GmbH
Ferdinand-Jühlke-Straße 7
99095 Erfurt, Deutschland
produktsicherheit@kolibri360.de